New Tools and Metrics for Evaluating Army Distributed Learning

Susan G. Straus, Michael G. Shanley,
Douglas Yeung, Jeff Rothenberg,
Elizabeth D. Steiner, Kristin J. Leuschner

Prepared for the United States Army
Approved for public release; distribution unlimited

ARROYO CENTER

The research described in this report was sponsored by the United States Army under Contract No. W74V8H-06-C-0001.

Library of Congress Cataloging-in-Publication Data

New tools and metrics for evaluating Army distributed learning / Susan G. Straus ... [et al.].
p. cm.
Includes bibliographical references.
ISBN 978-0-8330-5212-4 (pbk. : alk. paper)
1. The Army Distributed Learning Program (U.S.). 2. Military education—United States—Evaluation. 3. Distance education—United States—Evaluation. 4. Computer-assisted instruction—United States—Evaluation. I. Straus, Susan G.

U408.3.N268 2011
355.5—dc22

2011007452

Published 2011 by the RAND Corporation
1776 Main Street, P.O. Box 2138, Santa Monica, CA 90407-2138
1200 South Hayes Street, Arlington, VA 22202-5050
4570 Fifth Avenue, Suite 600, Pittsburgh, PA 15213-2665
RAND URL: http://www.rand.org/
To order RAND documents or to obtain additional information, contact
Distribution Services: Telephone: (310) 451-7002;
Fax: (310) 451-6915; Email: order@rand.org

Preface

Distributed learning (DL) is a key element of the Army's training strategy, and the Training and Doctrine Command has expansive goals for future use of DL. Careful assessment of new initiatives and pilots is important in order to support the case for needed resources and to choose options that best leverage available funding. However, the Army does not routinely assess the quality of DL training at the program level, and prior RAND research on specific training outcomes suggests that the Army is not realizing potential readiness benefits from its fielded DL courses.

This report describes a series of studies designed to develop and test tools and metrics to measure and understand training outcomes as well as to document the impact of Army DL courses at the program level. The research also assesses the capabilities of Army information technology systems to support efficient data collection for training evaluation. The report will be of interest to those involved in planning, developing, delivering, and evaluating distributed learning.

This research was sponsored by U.S. Army Training and Doctrine Command and was conducted within RAND Arroyo Center's Manpower and Training Program. RAND Arroyo Center, part of the RAND Corporation, is a federally funded research and development center sponsored by the United States Army.

The Project Unique Identification Code (PUIC) for the project that produced this document is ASPMO09186.

Correspondence regarding this report should be addressed to Susan Straus (sgstraus@rand.org).

For more information on the RAND Arroyo Center, contact the Director of Operations, Marcy Agmon (telephone 310-393-0411, extension 6419; fax 310-451-6952; email Marcy_Agmon@rand.org); or visit Arroyo's web site at http://www.rand.org/ard/.

Contents

Preface ... iii
Figures ... vii
Tables ... ix
Summary ... xi
Acknowledgments ... xxvii
List of Acronyms ... xxix

CHAPTER ONE
Introduction ... 1
Objectives of This Research ... 3
A Framework for Training Evaluation and Effectiveness ... 4
Organization of This Report ... 8

CHAPTER TWO
Surveys of Students' Experiences in DL ... 11
Survey Methods ... 14
Results ... 21
Future Administration of Surveys ... 44

CHAPTER THREE
Knowledge Retention of DL Material in the Phased Approach to Training ... 47
Challenges in Examining the Relationship Between Learning and Soldier Readiness ... 50
Assessment of Knowledge Retention from DL ... 51
Method ... 52

Findings 55
Conclusions and Recommendations for Improvement in DL Policies and Procedures 58

CHAPTER FOUR
Feasibility of Using Army Information Systems to Collect Training Evaluation Data 63
Method 64
Findings 65
Conclusions and Recommendations to Address Technical and Nontechnical Barriers 75

CHAPTER FIVE
Conclusion 85
Key Findings 86
The Suggested Way Ahead for Army Evaluation of DL 89
Final Thoughts 94

APPENDIX
A. Nongraduate Survey 95
B. Training Circumstances for Nongraduates 103
C. Graduate Survey 105
D. Graduate Survey Participant Characteristics 121
E. DL Training Circumstances for Graduates 123
F. Revised Nongraduate Survey 125
G. Revised Graduate Survey 131
H. Scoring Procedures for Student Surveys 143
I. Questions for Semi-Structured Interviews with SMEs About Army Information Systems 147
J. Service-Oriented Architecture 151

Bibliography 167

Figures

S.1. Reasons for Nongraduation (Nongraduate Survey) xiv
S.2. Graduates' Satisfaction with Aspects of DL xvi
1.1. Integrated Model of Training Evaluation and Training Effectiveness 5
2.1. Reasons for Nongraduation 22
2.2. Nongraduates: Factors Related to DL Program 25
2.3. Payment Status (Nongraduates and Graduates) 30
2.4. Ratings of Learning Preferences (Nongraduates and Graduates) 31
2.5. Graduates' Satisfaction with Different Aspects of DL 32
2.6. Graduates' Ratings of Course Length and Difficulty 33
2.7. Graduates' Satisfaction with Interaction 34
2.8. Types of Technical Problems Experienced by Graduates 36
2.9. Overall Satisfaction Ratings for Graduates by Orientation: DL and Classroom 40
3.1. Subsection of Alvarez, Salas, and Garofano (2004) Integrated Model Used to Guide Studies of Knowledge Retention 49
3.2. Model Tested in Assessments of Knowledge Retention 51
3.3. Distribution of Scores on Ordnance Knowledge Retention Test 56
3.4. Learning and Knowledge Retention in Ordnance Course 57
4.1. Mock-Up of a Training Dashboard Displaying Course-Level Data 80
5.1. Integrated Model of Training Evaluation and Training Effectiveness 91

Tables

1.1. Training Evaluation Measures 7
2.1. Types of Questions on the Nongraduate Survey................. 15
2.2. Courses and Graduation Rates: Nongraduate Survey........... 16
2.3. Types of Questions on the Graduate Survey 19
2.4. Courses Included in the Graduate Survey 20
2.5. Types and Quality of Technical Support Used by Graduates 37
5.1. Summary of Key Recommendations 88
H.1. Scoring Procedures for Nongraduate Survey 143
H.2. Scoring Procedures for Graduate Survey........................ 145

Summary

Distributed learning (DL) is a key element of the Army's training strategy, and the Training and Doctrine Command (TRADOC) has goals for expanding the future use of DL and changing how DL is developed and delivered. Although some individual proponent schools evaluate aspects of DL, the Army currently does not assess training outcomes of The Army Distributed Learning Program (TADLP) as a whole, i.e., across proponent schools. Program-level evaluation can play an essential role in the expansion of DL, in the identification of strategic directions for the DL program, and in ensuring the quality of training. First, careful assessment of new initiatives and pilots will help guide the choice of options that best leverage available funding. Second, evaluation will help the program compete for needed resources to support its expansion. TADLP resources for developing content have declined in recent years relative to other training accounts (Shanley et al., forthcoming); better documentation of the value of DL and its contributions to readiness could be essential to substantiate the case for increased funding. Third, evaluation can help identify specific areas for improvement in existing courseware content and delivery, technical matters, course management, training policy, and other aspects of DL design and implementation.

This report describes a series of studies designed to develop and test new tools and metrics to assess training and to document the impact of Army DL courses at the program level. The project on which this report is based builds on prior RAND research that has evaluated different aspects of Army DL. As a foundation for the current research, we used a model of training evaluation (Alvarez, Salas, and Garofano, 2004), which provides a broad conceptual framework to guide efforts to assess the quality of training and determine how to improve training programs. Below we list the components of the project:

- First, we developed and implemented two online surveys of students' experiences in DL: a *nongraduate survey*, focused on diagnosing why students fail to complete DL courses, and a *graduate survey*, which assessed reactions to completed courses. The surveys provide direct feedback from the ultimate customer (students), and are an important input for documenting DL's current contributions as well as for identifying the key improvements needed in courses and supporting systems.
- Second, we conducted assessments of *knowledge retention* of DL material and the association of learning and knowledge retention with individual soldier readiness for (i.e., performance in) follow-on resident training. Understanding the degree to which students learn from DL and retain knowledge as well as the extent to which DL influences soldiers' readiness for subsequent training is critical to documenting the value of DL and making improvements in future DL content and delivery.
- Third, we *assessed the capabilities of Army information systems* to facilitate enterprise-wide evaluation of DL outcomes. We approached this by interviewing Army subject matter experts (SMEs) concerned with information technology (IT) integration issues involving learning management systems (LMSs) and the Army Training Information System (ATIS). Assessing training effectiveness at the program level requires methods to collect and synthesize the data efficiently. Thus, a fundamental issue addressed in this study is how IT can be used to collect data to evaluate training and do so in a centralized, standardized way.

In the remainder of this summary, we highlight key findings from these three studies, along with specific recommendations relevant to each study. We conclude with recommendations to guide future Army evaluation efforts.

Surveys of Students' Experiences in DL

Understanding students' experiences in DL has received limited attention at the program level. For example, while individual proponent schools and centers assess student satisfaction with DL, such efforts are sporadic, and the use of different methods and metrics precludes integration of results (Straus et al., 2009). Prior research also shows that many key DL courses have low graduation rates, as indicated by the Army Training Requirements and Resources System (ATRRS), especially when compared to rates for resident courses (Shanley et al., forthcoming). However, the Army does not know why students do not begin courses after enrolling or why they fail to complete courses once they start.

Consequently, we developed two online surveys to assess student reactions in DL training courses, one focusing on nongraduates and one focusing on graduates. Each survey was pilot-tested in five high-priority Department of the Army (DA)-directed DL courses.[1] The nongraduate survey included 1,058 students (a 23 percent response rate from a larger sample), while the graduate survey included 431 students (a 30 percent response rate).

A Large Proportion of "Nongraduations" Were Due to Non-DL-Related Factors

We examined the reasons for nongraduations and categorized the results as shown in Figure S.1.

Results show that in 43 percent of the cases, students who appear as "nongraduates" in ATRRS had, in fact, already graduated or were taking the course for self-development, for which graduation is not required. Thus, in a large number of cases, students' nongraduation status was an artifact of record keeping, and graduation rates are higher than those derived from ATRRS. In addition, in 14 percent of the cases, nongraduation was attributable only to factors external to the DL program, such as mobilization or deployment or changes in occu-

---

[1] Most Army DL is currently part of a phased approach to training. Typically, a DL phase serves as a prerequisite for a resident phase of instruction. Some courses have multiple DL and resident phases.

Figure S.1
Reasons for Nongraduation (Nongraduate Survey)

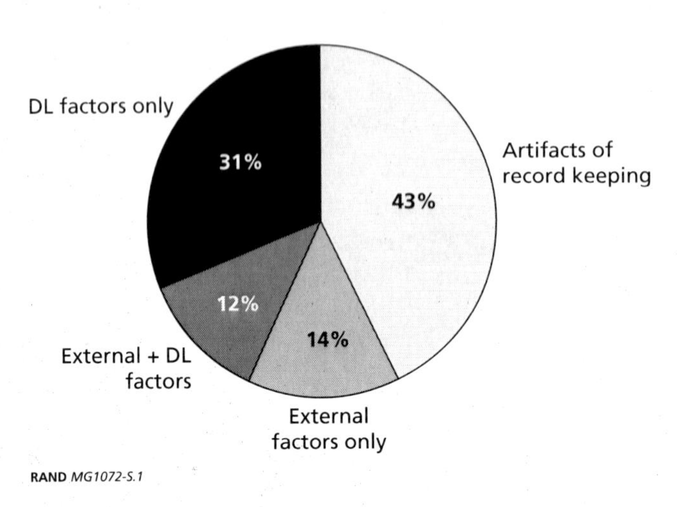

RAND *MG1072-S.1*

pations. Thus, over half the nongraduations (or cases that appear as such in ATRRS) cannot be attributed to DL courseware or policy.

The Main DL-Related Reasons for Nongraduation Were Related to Technical, Support, and Time Issues

Forty-three percent of nongraduations (DL factors only and external + DL factors in Figure S.1) could be tied to the DL program or its policies, either exclusively or in combination with external factors. The main DL-related reasons for nongraduations were technical problems, a lack of support, and insufficient time to complete coursework.

Two common technical issues for nongraduates involved lack of access to an Internet connection or to a reliable computer. The majority, however, had other technical difficulties. We did not further differentiate among other technical issues in this survey; however, responses to the end-of-course survey described below suggest the kinds of other technical problems that nongraduates may have experienced.

Issues with support centered on insufficient administrative and technical assistance. About 40 percent of these responses constituted cases in which students sought but did not receive help, primarily for technical issues. However, responses from the remaining 60 percent of these students indicated that simple administrative actions (e.g., follow-up notifications) would address their problems. For example, many who did not graduate were not aware that they had been enrolled in a DL course. This outcome points to issues with the existing policy of automatic enrollment in DL courses, suggesting the need for some follow-up.

Respondents with "time" issues cited a shortage of time to work on DL. Most of these students were also trying to complete the courses on personal time. Moreover, nongraduates were much more likely than graduates to use personal time to work on DL, suggesting that the amount of personal time used for required DL negatively affected graduation rates. Beyond the effect on graduation rates, this is a larger DL issue because DA policy states that students should be given duty hours (i.e., paid time) to complete required DL training.

Of equal importance are results indicating that some aspects of DL had a minimal effect on nongraduation rates. For example, issues related to DL courseware itself, such as course length or content, played a relatively minor role in explaining nongraduations. Furthermore, comparisons between nongraduates and graduates suggest that failure to complete courses cannot be attributed to an aversion to DL among students. Measures of student learning preferences show that while both nongraduates and graduates tended to prefer classroom learning to DL, the average rating for DL orientation (i.e., a preference for independent, self-paced, computer-based learning) was generally high among all students.

DL Graduates Expressed Moderate Satisfaction with Their Experiences in DL Courses, but Responses Also Suggest Areas for Improvement

Figure S.2 shows average ratings of graduates' satisfaction with multiple aspects of their experiences in DL courses. Three of the five measures in the figure are based on scales consisting of multiple items. On all five measures, the average rating was greater than 3 on a 5-point scale, suggesting that students were moderately satisfied with all aspects of the course that we assessed. The highest ratings (3.77) went to the technical support provided by Army and proponent school help desks.

Figure S.2
Graduates' Satisfaction with Aspects of DL

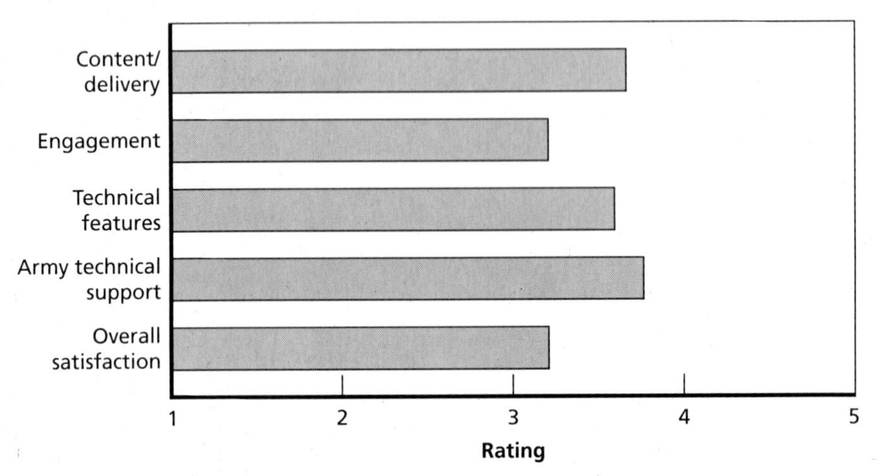

NOTE: Response options ranged from 1 = strongly disagree to 5 = strongly agree.
RAND *MG1072-S.2*

Students also gave generally favorable ratings to the quality of course content and delivery. However, ratings of the degree to which courses held the students' interest (engagement) were lower and suggest some need for improvement, especially because engagement was associated more strongly with overall satisfaction than were ratings of other aspects of DL.

Respondents were moderately satisfied with technical features of the courseware. However, 63 percent of respondents reported technical difficulties with the courseware, suggesting some need for improvement. The most common issues pertained to bandwidth or speed (e.g., delays in pages loading; difficulty playing audio or video files), followed by access to courseware (e.g., difficulty launching the course or receiving CD-ROMs). Fewer students reported problems navigating through courseware. Production quality (e.g., difficulty reading the text, excessively slow or fast narration) was the least problematic category of technical difficulties.

Issues Raised in Nongraduate and Graduate Surveys Suggest Recommendations for Change

The Army needs to capture information about student status in courses more accurately. Based on findings from the survey of nongraduates, we anticipate that modest changes to ATRRS and to administrative practices could result in significant improvements in DL outcomes:

- We recommend adding a field to ATRRS to document the student's purpose in enrolling in a course (i.e., requirement, self-development, refresher/reachback).
- The Army should also take steps to encourage updating of the ATRRS course graduation field for DL phases.
- Another option is to require that ATRRS have a valid graduation status for the DL phase in order for students to graduate from the resident phase of the course.
- There is also a need to enhance administrative support to ensure that students are aware they are enrolled in DL courses and to offer assistance when progress in the course appears lacking.

The Army should enforce the policy of paying soldiers for required training. The Army needs to enforce the policy allowing soldiers to use duty time for required training. One option is the current effort to create an "EDY" or educational duty status that students can use while working on required courses. Another option used by the Army National Guard (ARNG), observed in our studies of DL knowledge retention, is to allow students to come to the schoolhouse one week before the resident course to complete the DL prerequisite on paid time. This approach appeared to increase the likelihood that students would complete the DL phase and that they would do so prior to the start of the resident phase. Other options may also be needed. Absent an acceptable solution, the Army may want to move toward a particular type of blended-learning model (e.g., similar to the ARNG approach described above) wherein students complete the DL content while in residence.

The Army should seek to improve learner engagement and specific technical features in DL courseware. Based on the findings of the graduate survey, the Army should seek to increase interaction and to better engage learners in DL training. The Army's current move towards blended learning (which combines different modes of instruction, such as DL and resident training) may ameliorate some of these issues by making use of different media and varied methods of instruction and by providing more opportunities for interaction with instructors and peers.

The nature and frequency of graduates' responses indicate that improvements to technical features of DL courseware should focus on enhancing speed and access. Until bandwidth can be increased, one workaround (suggested by many of the students) is to provide low-bandwidth versions of courses or CD-ROMs for students in deployed locations or in other constrained settings.

The Army should administer surveys of students' experiences in DL on an ongoing basis. Our findings indicate that the nongraduate and graduate DL experience surveys are feasible and appropriate for ongoing use at the program level. The scales are psychometrically sound in terms of providing reliable measures, and results show a reasonable degree of variation in responses. From an administrative standpoint, the surveys are not burdensome to complete, and they are straightforward to interpret and score. The surveys can be adapted to address specific goals or topics of interest to TADLP or individual proponent schools.

We recommend that, moving forward, the Army use a common set of indicators on the surveys (with the option for schools to add questions to address local interests), as well as a common platform and software application. Use of both the graduate and nongraduate surveys should be part of routine quality improvement efforts.

Knowledge Retention of DL Material in the Phased Approach to Training

In the next part of this research, we conducted pilot studies to assess knowledge retention in two high-priority courses that use DL in the phased approach to training: Ordnance Mechanical Maintenance Basic Knowledge and Skills Course and the Battle Staff Noncommissioned Officer Course. The first course provided substantive results. The second course did not produce enough data to analyze the association between DL learning and knowledge retention, because some students did not complete the knowledge test (which was voluntary) and others had not completed the DL phase prior to resident training. However, results yielded important lessons for future studies with respect to student participation, DL policy, and Army information systems.

In the Ordnance course, we analyzed the level of knowledge retention following the DL phase; the effect of the lag following the DL phase on knowledge retention; and the associations of learning, retention, and readiness with organizational and student characteristics.

Knowledge Retention Was Associated with Time Spent on DL, Lag Time Between DL and the Resident Phase, Job Experience, and Cognitive Ability

We found, somewhat surprisingly, that the amount of time students worked on the DL phase was not associated with their performance on knowledge tests during DL, and performance in the DL phase was not associated with knowledge retention measured at the beginning of the resident phase. However, there was a positive correlation between time spent on DL and knowledge retention scores, indicating that students who spent more time working on the DL phase retained more knowledge. Greater knowledge retention was also associated with shorter lags between the time that students completed the DL phase and started the resident phase; greater relevant job experience (more civilian experience or longer time in their new military occupational specialty or MOS); and greater cognitive ability (as reflected in higher Armed Forces Qualification Test scores).

Recommendations to Address Issues Raised in the Knowledge Retention Assessment

Results of this analysis suggest directions for change in training policies and procedures to improve DL effectiveness:

- **Students should be encouraged to complete the DL phase of the course in a way that minimizes the time lag between the DL and resident phases.** In addition to changing the policy regarding when students can complete courses, the proponent schools should work to arrange training schedules to enable students to take the DL phase in close chronological proximity to the resident course but without "cramming" DL training into a short time period.
- **Participation in the DL phase may not be necessary for all students, particularly those with relevant experience in the subject area.** Data about relevant student characteristics could be used as a factor in determining whether some students can place out of all or part of the DL phase or be given a streamlined version of the DL course. Pre-tests could be used to accomplish the same goal.
- **DL courses might be more effective if students were supported in planning for sufficient time to complete the course.** We recommend sending students a "welcome" message upon enrollment that provides a timetable to progress through the course as well as contact information for further support. Instructors could also use system logs from the course LMS to identify and contact students who are not making steady progress.

Other changes could improve the Army's ability to evaluate knowledge and training performance. Rather than providing substantive results, the study in the Battle Staff course as well as our experience in the Ordnance course yielded a number of lessons about evaluating knowledge (or skill) retention and training performance:

- **Future studies should be conducted as part of ongoing Army quality improvement efforts** (like the process used in the Ord-

nance course) rather than as a research activity in which participation is voluntary.
- **Enforcing the policy of completing DL prerequisites in advance is also important** in order to study knowledge retention and performance in the phased approach to training.
- **Test evaluation is needed** to ensure that course tests are reliable, valid, and discriminate among high and low performers in the course.
- **Finally, a lack of interoperability among systems that contain relevant data poses a barrier to evaluating learning.** Those challenges may be somewhat less of an issue if such studies are conducted within the Army where student identifying information can be shared more easily.

Assessment of the Capabilities of Army Information Systems to Support Enterprise-Wide Evaluation

We conducted interviews with SMEs in TRADOC Headquarters, selected proponent schools, and Program Manager Distributed Learning Systems (PM DLS) who are concerned with IT integration issues involving LMSs and the Army Training Information System (ATIS). Our interviews with Army SMEs examined both technical and organizational factors that can affect data collection.

SMEs See Value in Evaluation Data

SMEs discussed student-level outcomes, such as performance on course tests, course usage statistics, and student reactions, and course characteristics, such as interactive multimedia instruction (IMI) levels, delivery modes, type of developer, and graduation rates. SMEs generally felt that gathering, sharing, and analyzing these types of data, if done well, could have significant value to schools, course developers, Army Training Support Center (ATSC), and possibly commanders and students. Most respondents felt that sharing data among schools might create a useful exchange of ideas and best practices for producing effective DL. A majority also felt that capturing student-level and course-level data could provide other benefits, including documenting the value of DL courses, justifying the need for resources, aligning programs of instruction with current doctrine, and providing data needed to better understand the effectiveness of online course development and delivery techniques.

However, There Are Numerous Technical Barriers to Data Collection, Sharing, and Analysis

System and usability problems, such as system crashes, are one source of technical issues that threaten the validity of training evaluation data as well as undermine the training process. However, the predominant technical issue threatening enterprise-wide training evaluation is poor interoperability among information systems. Evaluating training requires finding and querying multiple information systems, most of which are not connected to each other and so require specific expertise or authorization (or both) to access and use. Furthermore, identifying and joining data across systems can be problematic because each system may define and encode data in unique ways. Some systems are connected to each other using one-to-one ("pairwise") interfaces, but these connections are often difficult to maintain, particularly as more systems are connected. Other systems are not linked at all and require printing data from one system and manually reentering it into another.

Methodological, Organizational, and Policy Issues Further Impede Collection, Analysis, and Sharing of Training Evaluation Data

SMEs voiced concerns about the following issues:

- Aggregate measures of training quality are not comparable across schools and may be subject to misinterpretation.
- School staff might resist collecting data without a clear rationale.
- Measurement, data collection, and analysis could strain personnel resources and skills.
- The organizational culture is one in which schools generally consider performance measures to be their own concern.
- Results could be used to reduce schools' resources.

- Existing policies on how to collect, share, and use data are insufficient.
- Policies alone are not enough to encourage data sharing; top-down guidance and incentives are needed to overcome the tendency for schools to keep data to themselves.

Recommendations Based on the Assessment of Army Training Data and Information Systems

Technical impediments can be addressed in phases:

- **In the short term, the Army should help schools enhance their ability to evaluate training.** This process can be facilitated by moving to online administration of tests and surveys and by collecting different types of data within a single instrument. Manual studies should be continued until IT systems can fully support them; such studies not only provide results about the quality of training, but can yield lessons learned for designing automated collection of evaluation data.
- **In the medium term, the Army should build its ability to collect training evaluation data.** Critical to this ability will be the development of standards, which will help produce more complete, consistent, and available data. Standards are also necessary to move to service-oriented architecture (SOA). Modifications should also be made to allow training systems to collect data consistently. Web services should be modified to allow database queries.
- **In the long term, move to service-oriented architecture.** New architectures, notably SOA, have begun to emerge to support interoperability. SOA repackages the capabilities of systems into a set of "services" that can be located and used by other services, systems, or users on the network. Adopting SOA as the architecture appears to have the potential to greatly facilitate the collection and sharing of evaluation data, though not without associated risk and substantial cost. In principle, SOA would enable training systems to interoperate without prior agreement of any

kind, thereby bypassing the cost and lead time of creating specialized interfaces between systems.

Addressing nontechnical impediments to an enterprise-wide program of evaluation is a necessary component of an implementation plan:

- **Build end-user participation into all phases of process design.** We recommend inviting staff from proponent schools and centers as well as other organizations to participate in designing the processes used to collect and evaluate training data.
- **Establish the business case before requiring any new data collection.** Efforts should be undertaken to determine the value and feasibility of collecting various kinds of training evaluation data and to communicate the rationale for enterprise-wide evaluation efforts.
- **Develop appropriate policies to support evaluation.** The business case should be used to establish data-reporting requirements. These requirements should be crosswalked and integrated with relevant existing policies to avoid requiring overreporting. Policy should also be developed to spell out how training evaluation data will be used.
- **Provide requisite resources and incentives.** ATSC should ensure that the proponent schools and centers have the necessary resources to collect, analyze, and report evaluation data. This should include providing hardware and software for collecting, analyzing, and/or reporting data, resourcing the personnel needed to support these efforts, and providing training in analytical techniques.
- **Establish an entity to support enterprise-level training evaluation.** This office would provide analytical support to proponent schools and centers, identify relevant interoperability shortfalls and serve as a liaison to coordinate data exchange, coordinate data-collection efforts across the schools, integrate results (with input from schools) and report to ATSC and DA, and collect and disseminate lessons learned and best practices.

Recommended Way Ahead for Army Evaluation of DL

The results of this series of studies provide information about the current state of DL and suggest that a more comprehensive program of evaluation, better supported by the Army's IT systems, could provide major benefits to TADLP. Some of the current tools are ready to be turned over to the Army for implementation. Widespread adoption of the graduate and nongraduate surveys would help individual schools and centers to systematically evaluate DL courses and provide TADLP with aggregate measures of program quality. In addition, relatively little modification would be needed to use the surveys for other forms of DL, such as blended learning or mobile training teams (MTTs).

Evaluation of learning from DL needs additional investigation. Studies conducted in one or more large-scale DL conversion courses that provide both institutional and operational training (by serving as a job aid) could yield valuable lessons and raise awareness of the value of DL to the Army training community. We anticipate that these investigations will also set the stage for measuring the impact of training on job performance.

We also recommend evaluating a broader range of training outcomes. A critical area for future study is the effect of DL on acquisition and retention of skills (in addition to retention of knowledge). An important area for future investigation is developing and testing finer-grained measures of skills that distinguish among levels of student performance. Other outcomes to assess in a comprehensive evaluation program include:

- Post-training attitudes such as self-efficacy, which can be readily added to end-of-course surveys.
- Transfer performance, i.e., behavior on the job.
- Cost and benefits of DL.

In addition to expanding the range of measures used, training evaluation should be extended to other DL approaches. For example, the Army is moving toward increased use of blended learning, as described earlier, and mobile learning (mLearning), which involves the use of technologies such as netbooks, tablet computers, electronic book readers, personal digital assistants, and smart phones. Research efforts are needed to determine appropriate measures and methods for evaluating these forms of training.

In conclusion, improved tools and metrics for evaluating DL training can provide benefits to TADLP at multiple levels. At the student level, evaluation can enable training staff to determine student success and diagnose needs for remediation. At the course level, evaluation can show how DL affects learning and subsequent outcomes such as knowledge and skills retention and performance on the job; point to needs for improvement in course content or delivery; and determine the effect of interventions designed to enhance training quality or efficiency. At the program level, evaluation can demonstrate the value of DL and support the case for resources to meet program goals.

Acknowledgments

This project would not have been possible without the assistance of many people in the Army DL community. We wish to thank LTC Mark Lynch for his support and contributions to this project. We are grateful for the expertise, time, and effort from MSG David Wilkinson, SGM Jose Fragoso, 1SG Charles Mort, MSG Jason Leeworthy, and MSG DuJuan Warren from the U.S. Army Sergeants Major Academy; Ms. Belinda Ramirez, Dr. Dwayne Rogers, and Mr. John Goodwin from The Army Medical Department; Mr. Dan Oprish and Ms. Anne-Marie Warren from the Armor School and Center; Mr. David Nilsen from Alion Science and Technology; Ms. Cindy Major and Mr. Tom Littleton from the Maneuver Support Center; Mr. Timothy Ozman and Mr. John Deilus from the Ordnance School; Ms. Terry Hancock from the Financial Management School; Mr. Alvin Kahn from Army Training Support Center; and the many Army DL students who participated in this research. We also appreciate input on survey measures from SMEs from the Armor School and Center, Army Management Staff College, Army Medical Department, Army National Guard, Army Research Institute, Combined Arms Center–Center for Army Leadership, Maneuver Center of Excellence, Signal Center of Excellence, TRADOC, and U.S. Army Sergeants Major Academy. Many thanks to Christopher Paul from RAND and Eduardo Salas from University of Central Florida for their helpful reviews of this report. This study also benefited from the efforts of other RAND colleagues. We wish to thank Henry Leonard, James Crowley, Tom Bogdon, Rodger Madison, Scott Ashwood, Rachel Burns, and Gayle Stephenson for their contributions to this project.

List of Acronyms

AAR	After-Action Review
AC	Active Component
AFQT	Armed Forces Qualification Test
AJAX	Asynchronous JavaScript and XML
AKO	Army Knowledge Online
ALARACT	All Army Activities
ALMS	Army Learning Management System
AMEDD	Army Medical Department
ANCOC	Advanced Noncommissioned Officer Course (now Senior Leader's Course, or SLC)
API	Application Programming Interface
ARFORGEN	Army Force Generation
ARI	U.S. Army Research Institute
ARNG	Army National Guard
ASAT	Automated Systems Approach to Training
ASI	Additional Skill Identifier
ATIA	Army Training Information Architecture
ATIS	Army Training Information System
ATLD	Army Training and Leader Development

ATRRS	Army Training Requirements and Resources System
ATSC	Army Training Support Center
AUTOGEN	Automated Survey Generator
BCKS	Battle Command Knowledge System
BNCOC	Basic Noncommissioned Officer Course (now Advanced Leader's Course, or ALC)
C4ISR	Command, Control, Communications, Computers, Intelligence, Surveillance, and Reconnaissance
CAC	Common Access Card
CAD	Course Administrative Data
CBRN	Chemical, Biological, Radiological, and Nuclear
CBRNE	Chemical, Biological, Radiological, Nuclear, and Explosive
CCC	Captains Career Course
CID	Criminal Investigation Division
CMF(s)	Career Management Field(s)
CONUS	Continental United States
CTP	Common Training Picture
DA	Department of the Army
DL	Distributed Learning
DoD	Department of Defense
EDIPI	Electronic Data Interchange Personal Identifier
EDY	Educational Duty
ERP	Enterprise Resource Planning
ESB	Enterprise Service Bus
FM	Field Manual
GIG	Global Information Grid

GPA	Grade Point Average
HR	Human Resources
HQ	Headquarters
IA	Information Assurance
IMI	Interactive Multimedia Instruction
IT	Information Technology
LMS	Learning Management System
MANSCEN	Maneuver Support Center
MOS	Military Occupational Specialty
MTT	Mobile Training Team
NCO	Noncommissioned Officer
NCW	NetCentric Warfare
PC	Personal Computer
PII	Personally Identifiable Information
POI	Program Of Instruction
QI	Quality Improvement
RCCPDS	Reserve Components Common Personnel Data System
SaaS	Software as a Service
SGM	Sergeant Major
SME	Subject Matter Expert
SOA	Service-Oriented Architecture
SSN	Social Security Number
TADLP	The Army Distributed Learning Program
TAPDB	Total Army Personnel Database
TCM	TRADOC Capabilities Manager
TRADOC	Training and Doctrine Command

URL	Uniform Resource Locator
USASMA	United States Army Sergeants Major Academy
VTT	Video Teletraining
XML	Extensible Markup Language

CHAPTER ONE

Introduction

Distributed learning (DL) is a key element of the Army's training strategy, and the Training and Doctrine Command (TRADOC) has expansive goals for future use of DL. While DL is currently used for a small portion of a relatively small number of courses, in the future it is expected to be a much more important and widely used training resource for the Total Force (TRADOC, 2011). The TRADOC Capabilities Manager for DL (TCM-DL) also has plans to change how DL is developed and delivered, e.g., by moving from a client-server paradigm with training delivered on desktop computers to "cloud computing" (i.e., services provided via the Internet) with mobile learning (mLearning) training devices, and by shifting from long courses with protracted development time frames to "chunked content" that can be developed rapidly.

There are many promising proposals for expanding the DL program to achieve the Army's goals, and careful assessment of new initiatives and pilots is important to ensure that the Army chooses the options that best leverage available funding. In addition, an increase in funding will be needed to support the Army's vision for DL. The Army Distributed Learning Program (TADLP) resources for developing content have declined in recent years relative to other training accounts (Shanley et al., forthcoming); thus, better documentation of the value of DL could be essential to support the case for increased funding.

Although some individual proponent schools have conducted evaluations of aspects of DL, TADLP does not routinely evaluate DL across proponent schools, i.e., *at the program level*, which is needed to understand the effectiveness of DL within Army training as a whole.[1] Furthermore, a RAND Arroyo Center study of student enrollment and completion rates across TADLP (Shanley et al., forthcming) found low utilization of fielded courses, indicating that the Army is not realizing potential readiness benefits of DL. However, the Army does not know why students are failing to enroll or why they are dropping out, because TADLP does not systematically collect data on student experiences with DL.

Program-level evaluation also is important to measure the operational impact of DL in terms of student learning. Understanding DL's contribution to learning is important both to support the case for DL and to understand how learning influences soldier readiness for subsequent training and performance on the job. By monitoring learning outcomes, the Army can ensure that changes to courses have the intended results (e.g., knowledge enhancement, skill acquisition) and that initiatives to increase training efficiency do not have unintended negative consequences.

In sum, documenting the impact of DL across proponent schools will contribute to several important TADLP goals. Evaluation of training outcomes is needed to determine the program's current effectiveness; to help the program better compete for needed resources; and to identify specific areas for improvement in courseware content and delivery, technical issues, course management, training policy, or other aspects of DL design and implementation. Assessment will help the Army identify strategic directions for the DL program. Demonstrating DL's contribution to readiness will instill confidence in the program and help cultivate awareness of the value of DL among stakeholders—from students to decision makers in the Department of the Army (DA).

This report describes a series of studies designed to develop and test new online tools and metrics to measure and understand training outcomes and to document the impact of Army DL courses at the pro-

[1] We distinguish efforts at individual proponent schools, which are decentralized, from efforts at the program level, which would assess training effectiveness across proponent schools.

gram level. This project also assessed the capabilities of Army information systems to support efficient data collection, including the integration of collection efforts with the Army Training Information System (ATIS). These research activities address a need identified by Salas and Cannon-Bowers (2001) in their discussion of evaluating learning outcomes from training: "The next frontier and greatest challenge in this area is in designing, developing, and testing online assessments of learning and performance" (p. 487).

Objectives of This Research

Prior RAND Arroyo Center research has examined course usage statistics (Shanley et al., forthcoming) and conducted independent evaluations of course content and delivery (Straus et al., 2009). In the research reported here, we developed and tested additional measures to assess student reactions to DL and cognitive learning from DL. We also collected data from subject matter experts (SMEs) regarding use of Army information systems for collecting training evaluation data. More specifically:

- We developed and pilot-tested two online surveys to assess student experiences with DL: one that diagnoses reasons for students' failure to complete DL courses (i.e., the nongraduate survey), and another that assesses graduates' reactions to DL courses (i.e., the graduate survey).
- We pilot-tested two assessments of knowledge retention to measure DL's contribution to students' cognitive learning and readiness for subsequent training.
- We conducted interviews with SMEs to assess the capabilities of Army information systems for ongoing measurement of DL effectiveness.
- We identified short-term and long-term actions to enhance evaluation of DL within TADLP, improve the quality of DL courseware and training processes, and increase DL utilization.

In the following section we describe the overarching framework that guided these efforts. The specific method used for each task is described in subsequent chapters.

A Framework for Training Evaluation and Effectiveness

As a framework for this series of studies, we adapted a model of training evaluation and training effectiveness proposed by Alvarez, Salas, and Garofano (2004) (see Figure 1.1). The model focuses on training as *a learning system,* which encompasses measures of learning and performance outcomes (which Alvarez, Salas, and Garofano refer to as *training evaluation*) as well as factors that influence these outcomes (referred to as *training effectiveness*). *Training evaluation* is important because it indicates whether a training program meets its intended goals. For example, training evaluation might show how satisfied trainees are with a course or whether the skills learned in training transfer to the job. *Training effectiveness* focuses on understanding the reasons for training outcomes, e.g., why students are more or less satisfied with a course or why training does or does not affect subsequent job performance. Assessing the reasons for training outcomes enables designers, developers, and instructors to modify training in order to improve quality.

We use this model because it is well grounded in training theory and research and because it provides a broad conceptual framework to guide efforts to assess the quality of training and determine how to improve training programs. The model integrates constructs from a number of other influential models (Holton, 1996; Kraiger, 2002; Tannenbaum, Cannon-Bowers, Salas, and Mathieu, 1993; cf. Alvarez, Salas, and Garofano, 2004). It is based on a comprehensive review of research findings from 73 studies of adult training published during the 10 years prior to development of the model.

The integrated model consists of four levels, as shown in Figure 1.1. Shaded boxes denote constructs that we have added to the model.

Level 1: Needs Analysis. The first level consists of needs analysis, which determines what knowledge and skills should be addressed in training. Needs analysis is included in the model because it influences

Figure 1.1
Integrated Model of Training Evaluation and Training Effectiveness

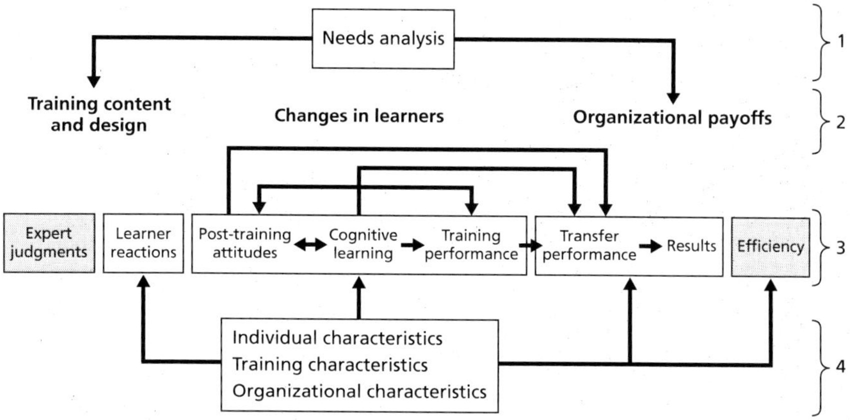

Needs analysis
1
Training content and design
Changes in learners
Organizational payoffs
2
Expert judgments
Learner reactions
Post-training attitudes
Cognitive learning
Training performance
Transfer performance
Results
Efficiency
3
Individual characteristics
Training characteristics
Organizational characteristics
4

SOURCE: Adapted from Alvarez, Salas, and Garofano, 2004, p. 393, Figure 1. Used with permission.
RAND *MG1072-1.1*

training evaluation. However, we do not address needs analysis in this study.

Level 2: Target Areas of Evaluation. The second level depicts three target areas of training evaluation: content and design of training (which includes delivery and validity of training), changes in learners as a result of training, and effects on the organization due to training.

Level 3: Measures to Assess Target Areas. The third level consists of training evaluation measures. These measures build on Kirkpatrick's (1994) four-level classification of training outcomes, which consists of learner reactions, learning, behavior (on-the-job performance), and results. Alvarez, Salas, and Garofano (2004) propose that training content and design can be measured by reactions to training (but not by the other measures in the model). We also believe that training content and design can be evaluated by experts (Straus et al., 2009), and so have added that to the figure. In addition, changes in learners can be measured by post-training attitudes, cognitive learning, and training performance. Payoffs to the organization can be measured by transfer performance and results. In addition, payoffs to the organizations occur in the form of a variety of efficiency measures of training, such as number or proportion of personnel trained, time it takes to achieve course standards, content development cost and cycle time, and travel and accommodation costs (Paradise and Patel, 2009; Shanley et al., forthcoming). We have thus added "efficiency" to the model. Definitions of the training evaluation measures from the model as related to Army training are shown in Table 1.1.

Level 4: Characteristics That Affect Training Evaluation Outcomes. The fourth level in the model consists of individual, training, and organizational characteristics that affect training evaluation outcomes. Alvarez, Salas, and Garofano referred to the measure of these variables as "training effectiveness" but we use the term "explanatory variables," as "training effectiveness" is often interpreted more broadly to mean overall training quality. Examples of individual explanatory characteristics include abilities, demographics, experience, motivation, and personality traits. Training characteristics refer to factors such as instructional style, practice, and feedback. In DL, additional training characteristics might include delivery mode and level of interactivity in interactive multimedia instruction (IMI). Organizational characteristics include factors such as the climate for learning, training policies, and administrative procedures (e.g., registration for courses). Individual, training, and organizational characteristics affect one or more of the evaluation measures. Although Alvarez, Salas, and Garofano hypothesize that these characteristics influence only certain measures (e.g., that only individual characteristics affect learner reactions), we hypothesize that individual, training, and organizational characteristics can affect most of the measures in the model.[2]

The arrows within and between levels show how various constructs are associated with one another. For example, Alvarez, Salas, and Garofano postulate that post-training attitudes, cognitive learning, and training performance influence transfer performance, which in turn affects results. Post-training attitudes are presumed to have

---

[2] For example, organizational policies that provide duty hours to complete DL might positively influence student reactions, cognitive learning, or course completion rates by providing more time to work on training.

Table 1.1
Training Evaluation Measures

Target Area	Measure	Definition	Method Commonly Used
Training content and design	Learner reactions	Individuals' attitudes toward aspects of training such as usefulness and relevance of content and quality of instruction	End-of-course surveys.
	Expert judgments	Evaluations of training content and delivery	Army course validation process. See also Straus et al. (2009).
Changes in learners	Post-training attitudes	Affective outcomes such as self-efficacy, motivation, and attitudes toward the training objectives	End-of-course surveys.
	Cognitive learning	Acquisition of knowledge	Paper-and-pencil or online tests. Can be used to measure immediate learning or delayed learning (Alliger et al., 1997) i.e., knowledge retention.
	Training performance	Acquisition of skills	Observable demonstrations of skills (skills tests).
Payoffs to organization	Transfer performance	Behavioral changes on the job as a result of training	Supervisors' evaluations or objective performance measures (e.g., error rates, time to completion).
		Performance in subsequent training	Cognitive learning or training performance.
	Results	Changes in outcomes due to behavior changes	Quality or quantity of unit performance.
	Efficiency	Optimal use of resources	Administrative records data.

reciprocal effects with cognitive learning and training performance in light of research findings showing that the direction of the relationship can go either way (i.e., students with more favorable attitudes learn more, and students who learn more have more favorable attitudes). Students with more experience relevant to the course or higher motivation (individual characteristics) may show higher levels of cognitive learning, training performance, and transfer performance.

Prior RAND Arroyo Center research on Army DL has addressed course content and design via use of expert judgments (Straus et al., 2009) and efficiency in terms of course utilization rates and course development cycle time (Shanley et al., forthcoming). This study is concerned with two other measures described in Table 1.1: learner reactions and cognitive learning. We focused on end-of-course reactions because the Army currently does not measure students' attitudes toward DL in a way that provides information at the program level. We also measured student reactions to diagnose reasons for low graduation rates, which is characteristic of most DL courses. We focused on cognitive learning in terms of declarative knowledge because it is a fundamental aim of DL (especially IMI) and is necessary for acquisition of higher-ordered knowledge and skills. In addition, we measured a variety of individual and/or organizational-level explanatory variables associated with reactions and learning that are described in subsequent chapters.

Organization of This Report

The remainder of this report is organized as follows:

- In Chapter Two we describe development, pilot testing, and results of two surveys of students' experiences in DL: one of students who did not complete DL courses (nongraduate survey), and one of DL graduates (graduate survey).
- In Chapter Three we present results of pilot studies of knowledge retention, which sought to measure the extent to which students learn and retain knowledge from DL courses and the impact of DL on subsequent residential training.
- In Chapter Four we describe the results of our interviews with SMEs regarding the capabilities of Army information systems to collect, analyze, and disseminate data on the quality of DL.

- In Chapter Five we summarize our findings and discuss the implications for policy. We also describe topics for future research.
- The appendixes provide supporting material. Appendixes A through E provide the survey questions, statistics describing survey participants, and the circumstances in which they participated in training. Appendixes F and G present revised versions of the surveys, and Appendix H explains how to score survey responses. Appendix I contains the questions used in the interviews of SMEs, and Appendix J provides a technical description of service-oriented architecture.

CHAPTER TWO

Surveys of Students' Experiences in DL

Student reactions are an important part of assessing the quality of training. Surveys of trainees can be used to measure reactions to numerous aspects of training, including the quality of course content, design, and delivery; instructor style; and organizational support. In particular, a recent meta-analysis by Sitzmann, Brown, Casper, Ely, and Zimmerman (2008) showed that trainee reactions are strongly associated with situational characteristics such as instructor style and opportunities for interaction with instructors and peers. Thus, post-training surveys can provide valuable information that can be used to improve course quality.

Student reactions also reflect other training evaluation outcomes. In a meta-analysis of the associations among training evaluation measures, Alliger and his colleagues found that reactions regarding the usefulness of training were positively associated with learning and transfer performance (Alliger, Tannenbaum, Bennett, Traver, and Shotland, 1997). More recently, Sitzmann et al. (2008) found that reactions to training were positively associated with immediate (post-training) declarative and procedural knowledge (although not with delayed procedural knowledge).[1] (See Chapter Three for a more in-depth dis-

[1] Sitzmann et al. (2008) found that post-training self-efficacy, i.e., confidence in one's ability to perform a task, was associated with delayed procedural knowledge and that it was more strongly associated with other learning outcomes than were measures of reactions. However, measures of self-efficacy are typically linked to specific course objectives; therefore, we did not include self-efficacy in the end-of-course survey in the current study, as this survey is intended to be applicable across a range of courses. As discussed in Chapter Five of this report, we recommend that proponent schools tailor the surveys to their courses and include measures of self-efficacy.

cussion of learning.) Others (Brown, 2005; Kraiger, 2002) argue that trainee reactions are associated with organizational outcomes such as enrollment in future training and attrition rates, although there are few studies of the relationship between reactions and these outcomes (Sitzmann et al., 2008).

Post-training as well as pre-training surveys can also be used to gather data on affective learning outcomes (e.g., self-efficacy, motivation, and attitudes toward training topics) and trainee individual characteristics (e.g., anxiety, self-efficacy, motivation, attitudes, and knowledge; see Table 1.1). Analyzing changes in affective outcomes, such as differences in self-efficacy and attitudes before and after training, can provide one indicator of training effectiveness.[2] When administered prior to training, measures of individual characteristics like anxiety can be used to identify students who would benefit from interventions such as added instructor support (Sitzmann et al., 2008). Finally, reactions to training predict changes in attitudes, whereas measures of learning do not. That is, trainees who have more positive reactions to training will be more receptive to attitude change. Thus, when training is intended to change attitudes (rather than to build skills), as in sexual harassment, ethics, or diversity training, it is particularly important to measure student reactions to training and to design courses in ways that foster trainee satisfaction (Sitzmann et al., 2008).

The results of Sitzmann et al.'s analysis are especially relevant for DL. The authors found that the associations of reactions and outcomes, including post-training motivation, self-efficacy, and declarative knowledge, were stronger in courses that used a high rather than a low level of technology. These results indicate the importance of assessing student reactions in technology-mediated instruction.

In the present analysis, we developed two surveys of student experiences in DL, one to diagnose reasons for high attrition from DL training, and another to help determine the quality of existing

---

[2] "Pre" versus "post" comparisons cannot be used to draw conclusions about the effect of training on outcomes. An appropriate experimental design is needed to rule out alternative explanations for the outcomes and conclude that training accounts for any change in outcomes.

DL courses and to improve the implementation and quality of future courses. Motivation for the first survey came from the fact that many key DL courses show low graduation rates in Army databases especially when compared to rates for resident courses (Shanley et al., 2008). An important policy question for TADLP is whether those rates could be increased to expand the impact of the program and improve return on investment. To address this question, we designed and pilot-tested a survey of DL "nongraduates" to help explain why students fail to complete DL courses or fail to begin them after enrolling.

Motivation for the second survey came from the fact that across all DL courses, understanding students' experiences in DL has received limited attention at the program level. Although some individual proponent schools and centers assess student satisfaction with DL, such efforts are sporadic, and the use of different methods and metrics precludes the integration of results (Straus et al., 2009). After action reviews (AARs) in particular, while potentially providing rich qualitative information, are not conducive to producing systematic or quantifiable data. Consequently, we designed and pilot tested an end-of-course survey of students' reactions to DL. The survey focuses on IMI but could be adapted for other modes of DL.

These surveys are intended to be applicable to a wide range of courses and to provide a foundation for ongoing evaluation of DL courses across Army schools. Our implementation of the surveys demonstrates the kinds of data that the Army could collect on a regular basis, identifies lessons about how to best implement the surveys, and points to potential directions for improvement within TADLP. Ultimately, these surveys can be used to increase utilization of DL, enhance its quality, and demonstrate its value to stakeholders.

This chapter describes survey development, implementation, and results. We first present methods for both surveys. We then present findings from the surveys and make recommendations for improving DL courseware and supporting systems. Finally, we discuss considerations for conducting these surveys on a regular basis. We present revisions of the surveys for ongoing use and provide procedures for scoring survey responses in Appendix F and Appendix G.

Survey Methods

Nongraduate Survey

Development of Survey Questions. The nongraduate survey sought to identify the reasons that students do not complete DL courses or do not begin courses after enrolling. Survey questions were developed based on: (1) interviews with DL proponent school staff and administrators of the Army Training Requirements and Resources System (ATRRS), the official Army database for recording training enrollments and graduates (Shanley et al., forthcoming); and (2) RAND staff experience with Army DL courseware (Straus et al., 2009). These items were vetted with five SMEs from the Army DL community and revised based on their feedback. We combined these questions with items also used in the graduate survey (described later in this chapter), including organizational and training characteristics (i.e., questions about training policy and circumstances in which students took the course) and individual characteristics (demographic questions and individual learning preferences).[3]

The nongraduate survey consisted of questions in the categories shown in Table 2.1. See Appendix A for a complete list of survey questions.

We programmed the survey using SelectSurvey, a commercially available Web-based tool developed by ClassApps. The survey was hosted on a secure RAND server. We used conditional logic so that questions varied by students' course completion status. Participants who responded that they had enrolled but had not completed the course were asked the full set of survey questions. Participants who enrolled but did not begin the course were asked all questions except for items about training circumstances. Participants who were unaware they were enrolled, who had completed the course, or who were still working on the course were asked no further questions. Consequently, the number of responses per question varies.

---

[3] Development and vetting of items measuring training policy and circumstances, demographic characteristics, and learner preferences are described in later in this chapter.

Table 2.1
Types of Questions on the Nongraduate Survey

Category	Items (see Appendix A)	Description
Course and course status	1–2	Course number/name Course completion status
Reasons for enrolling but not starting a course or starting but not completing a course	3 and 4, respectively	Students were asked to check all that apply from a list of options for each question. Alternative reasons could also be entered in an open-ended format.
Training policy and circumstances	5–10, 13	Conditions in which the student took the course, e.g., Location Amount of time spent working on the course Payment status Media used (Internet and/or CD-ROM)
Demographic characteristics	11–12	Military component Grade or rank
Individual preferences	14–15	DL and classroom orientation Attitudes toward using information technology (Agarwal and Prasad, 1998)

Recruitment Procedures. For the pilot implementation of the nongraduate survey, we selected five large, high-priority DA-directed DL courses which, according to ATRRS, had high noncompletion rates in FY08 (see Table 2.2). The courses selected had a collective graduation rate of 62 percent. By way of comparison, resident courses typically have graduation rates exceeding 90 percent.

All five of these courses used the phased approach to training, in which a DL phase using IMI serves as a prerequisite for a resident phase.

To generate the pool of potential participants for the survey, we used ATRRS to identify students who had enrolled in a DL course in FY09 but did not record a graduation or enroll in follow-on resident phases by the end of October FY09.[4] We contacted students by email

---

[4] Although we based selection of courses for the survey on FY08 outcomes, we recruited participants from FY09 in order to represent the most recent DL experiences. The survey

Table 2.2
Courses and Graduation Rates: Nongraduate Survey

DL Course	FY08 Enrollees	FY08 Graduations	FY08 Grad Rate
CBRN Specialist (74D10)	627	245	39%
Motor Transport Operator ALC (88M30)	1,483	834	56%
CMF 63/94 ANCOC	1,170	739	63%
First Sergeant Course	4,611	3,174	69%
AMEDD Captains Career Course (CCC)	1,940	1,110	57%

NOTE: CBRN means "Chemical, Biological, Radiological, and Nuclear." A newer term is CBRNE, meaning "Chemical, Biological, Radiological, Nuclear, and Explosive." ALC means "Advanced Leader's Course," formerly called "Basic Noncommissioned Officer Course" (BNCOC). ANCOC means "Advanced Noncommissioned Officer Course," now called the "Senior Leader's Course" (SLC). CMF means "Career Management Field." AMEDD means "Army Medical Department."

to request their participation in the survey, followed by a reminder message sent after 1–2 weeks. Participation in the survey was anonymous in order to protect student identity and to encourage open and honest responses.

Participants. We successfully contacted 4,629 students across the five courses (after adjusting for undeliverable email messages). A total of 1,058 students participated in the survey, yielding a response rate of 23 percent. This rate of participation is commensurate with other Web surveys (Schonlau, Fricker, and Elliot, 2002). Active Component (AC) soldiers accounted for 52 percent of the participants and were a substantive presence in all five courses, even courses originally designed with the needs of the Reserve Components (RC) in mind. The majority of participants were in grades E4–E8 and O2–O4. Appendix B reports the circumstances under which students participated in training.

A number of students responded to the survey invitation through return email rather than by logging into the survey site. The responses allowed us to identify students who, in fact, had completed the course, or who had some special circumstances that disquali-

included a question that allowed us to exclude students from the analyses who were still making efforts toward completion.

fied them from our participant pool (e.g., the email address associated with the student's record in ATRRS was for a different person). Other email messages described reasons for nongraduation; we used this information in our analysis of survey results, although we lacked other data about these students (e.g., student background and demographic characteristics).

Of the 1,058 who responded to our survey (either by taking the survey or responding through email), 656 participants, or 62 percent of the total, were included in the analysis. We dropped 388 students because they were still actively working on the course. An additional 14 were dropped because there was insufficient information to determine the reason for nongraduation.

Although not part of the survey, an additional 350 students were included in the analysis based on ATRRS records; these were students who had enrolled in both the DL courses and the corresponding follow-on resident courses, but who were shown in ATRRS as nongraduates in the DL courses. These students, representing about 10 percent of the original list of nongraduates, were not surveyed because we assumed that their enrollment in the resident phase meant that they had completed the DL phase.[5] They were included in the larger analysis as instances where ATRRS failed to reflect the true graduation status.

Graduate Survey

Development of Survey Questions. To develop items for the graduate survey, we first reviewed the literature to identify established scales used to measure student reactions to DL training. We created a pool of 91 items drawn from the following sources:

- Army surveys and AARs: TRADOC (2004); Morey, Bush, Beebe, McPhail, and Bickley (2009).

---

[5] We made this assumption because the resident phase requires graduation from the DL course. Results from our survey of graduates (described later in this chapter) suggest, however, that about 10 percent of these students completed the DL course during or after the resident stage rather than before it.

- Other published surveys of Web-based or classroom training. These included Bernard et al. (2004); Fisher, Wasserman, and Orvis (2004); Hamilton, Klein, and Lorie (2000); Lewis and Seymour (undated); Lawther and Walker (2001); Peterson's Distance Learning Assessment; Online Learning Resource survey (University of Wisconsin-Whitewater, 2006); and Wang, Dziuban, and Moskal (2000).
- Previous RAND Arroyo Center research evaluating Army interactive multimedia instruction (IMI) courseware (Straus et al., 2009), which was based on checklists from the training development community.

The item pool consisted of close-ended items comprising the following categories: (1) training policy and circumstances; (2) demographics characteristics; (3) technical features of the course; (4) audio-visual features of the course; (5) quality of instruction; and (6) classroom and DL orientations. The item pool also included several open-ended questions.

We sent the item pool to 26 SMEs in the Army DL community, including training division, branch, or program chiefs; course managers; instructional systems specialists; training developers; and students. SMEs were asked to rate each item based on its importance (from 1 = not at all important to 5 = very important) and to comment on how clear and understandable the items were. SMEs suggested revisions and proposed additional survey topics or items. Twelve SMEs (46 percent) provided feedback on the items, which we used to revise survey items. As with the nongraduate survey, we programmed the graduate survey using SelectSurvey and hosted it on a secure RAND server.

The graduate survey consisted of questions in the categories shown in Table 2.3. The full survey is provided in Appendix C.

Response options for most items were 5-point scales ranging from 1 = strongly disagree to 5 = strongly agree. Other questions had yes/no options or were open-ended. As with the nongraduate survey, we created conditional logic so that questions not relevant for an individual could be skipped depending on his or her earlier answers. As a result, sample sizes vary across questions.

Table 2.3
Types of Questions on the Graduate Survey

Category	Items (see Appendix C)	Description
Course and course status	1–5	Course number/name Course completion status
Training policy and circumstances	6–13, 18	Conditions in which the student took the course, e.g., Location Amount of time spent working on the course Payment status Media used (Internet and/or CD-ROM)
Demographic characteristics	14–17	Military component Grade or rank Experience in MOS (if applicable)
Technical features, user interface, and support	19–28	Types of technical difficulties encountered Types of technical support used and satisfaction with support Satisfaction with resources that provide support for course content (e.g., field manuals [FMs] and glossaries)
Course content and delivery	29–45	Quality of instruction Relevance of course content Quality of opportunities for practice Engagement, i.e., degree to which the course held the student's interest Amount and quality of interaction with instructors and other students
Overall satisfaction	46–48, 51	Overall level of satisfaction with the course and willingness to take additional DL courses
Individual preferences	49–50	DL and classroom orientation Attitudes toward using information technology (Agarwal and Prasad, 1998)

Recruitment Procedures. We selected a convenience sample of courses for the pilot implementation of the graduate survey after discussion with program chiefs and course managers in several schools. All courses were high-priority DA-directed training, and, as in the

Table 2.4
Courses Included in the Graduate Survey

Proponent	Course	Number of DL Hours	Approx. Number Students/Year
Armor	Cavalry Scout, 19D10	75	500[a]
	Armor Crewman, 19K10	75	135[a]
USASMA	Battle Staff NCO	75	2,000
AMEDD	Captains Career Course	107	1,000
MANSCEN	Criminal Investigations (CID) Special Agent, 31D20/30, Phase I	136	30
	CID Special Agent, 31D20/30, Phase III	101	30

NOTE: USASMA is the U.S. Army Sergeants Major Academy; MANSCEN is the Maneuver Support Center.
[a] FY09.

survey of nongraduates, all were prerequisite phases for subsequent resident training (see Table 2.4).[6]

We used several procedures to recruit students for participation in the survey. In several Battle Staff Noncommissioned Officer (NCO) classes, instructors asked students to complete the survey at the beginning of resident training (the DL phase is a prerequisite for the resident phase). Time was set aside for students to complete the survey. In the AMEDD CCC, we identified students in ATRRS with FY09 course completion dates and contacted these students by email. In the Armor and MANSCEN schools, a link to the survey was added to the course homepage in BlackBoard with a request to participate in the survey after completing the DL phase of the course. In addition, for the two Armor School courses, we identified students in ATRRS who completed the courses in 2009 and contacted them by email.

Students were provided with a URL for the survey. As with the nongraduates survey, participation was anonymous in order to protect student identity and encourage open and honest responses.

---

[6] In addition to the courses listed, a link to the survey was added to the BlackBoard course home pages for students in Ordnance 91/63A, B, D, H, M and MANSCEN 21E, H, K, R, N, T, and W courses. Only one student from each of these proponent schools participated in the survey, so we excluded their data from the analyses.

Participants. Four hundred sixty-five students participated in the graduate survey. Incomplete cases and other unusable responses were eliminated, leaving a total of 431 responses, with an overall response rate of 30 percent.[7] Response rates for the individual courses ranged from 13 percent to 96 percent and varied depending on the course and method of recruitment for the survey.[8] Student demographic characteristics are reported in Appendix D.

The vast majority of students completed the course in three months or less, and most (89 percent) reported that they had sufficient time to complete the course. Although there was some variation across courses, participants worked on the course primarily at home or at a CONUS military facility. With the exception of the Battle Staff NCO course, students generally took the course online on a high-speed Internet connection. Appendix E presents detailed training circumstances for participants.

Results

We now present results for both the nongraduate and graduate surveys. We begin by discussing factors that explain low graduation rates for some courses based on results from the survey of nongraduates. Next, we explore further possible reasons for nongraduation by comparing differences in explanatory variables (use of personal versus paid time and learning preferences) for both nongraduates and graduates. We then present results from the graduate survey, describing student satisfaction with different aspects of their experience in DL.

---

[7] These include 25 cases in which participants completed only the background questions and did not respond to questions about their experiences in the course; in addition, two participants indicated that they quit the DL course and therefore were not asked to answer any additional questions. We also eliminated responses from other courses for which we had only one or two respondents.

[8] In addition to the courses listed in Table 2.1, a link to the survey was added to the Black-Board course home pages for students in Ordnance 91/63A, B, D, H, M and MANSCEN 21E, H, K, R, N, T, and W courses. Only one student from each of these proponent schools participated in the survey, so we excluded their data from the analyses.

A Large Proportion of "Nongraduations" Were Due to Non-DL-Related Factors

The first goal of the nongraduate survey was to explain why enrollees in DL courses show a "nongraduation" status. Equally important, we sought to use the survey to aid the training community in determining the extent to which the incidence of nongraduations suggests a need for changes in DL policy, approaches, or processes.

Our findings indicate that there are a number of reasons why students may fail to show a graduation status in ATRRS. Based on the survey responses, we categorized reasons for nongraduation in ATRRS as follows (see Figure 2.1):

- **Artifact of recordkeeping:** Student completed the DL course, but this is not shown in ATRRS. In the phased approach to training, ATRRS requires entry of a completion date only for the resident phase, not for the DL phase. It is also possible that a student enrolled in the DL course for reasons other than graduation (e.g., self-development) and did not seek documentation of course completion in ATRRS.

Figure 2.1
Reasons for Nongraduation

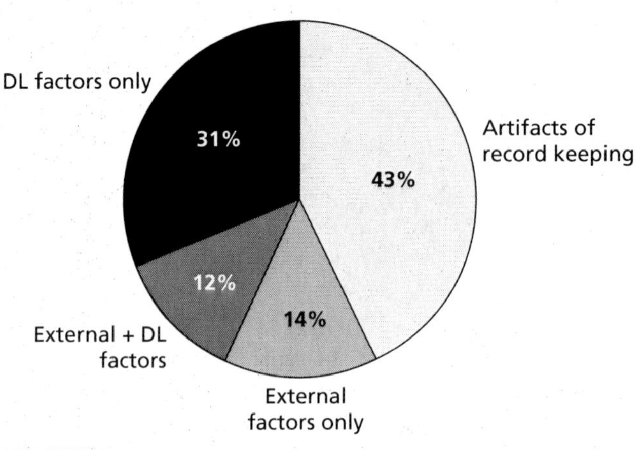

RAND *MG1072-2.1*

- **Reasons external to the DL course itself:** Student's personal circumstances changed, preventing completion of the course (e.g., the student left the Army, changed occupations, had a personal emergency, or was deployed). In ATRRS, these circumstances often result in withdrawal from or a "no-show" for residential courses, rather than a nongraduation. For residential courses, an enrollee is one who shows up at the training location. In contrast, for DL courses, an enrollee is one who has been registered for the course. Thus, graduation rates for DL courses are not strictly comparable to those for resident courses and would be expected to be somewhat lower due to the different ways in which enrollment is defined.
- **Reasons associated with the DL itself:** Some experience specific to the DL course prevented or deterred the student from completing the course (e.g., technical problems, issues with course content, problems getting support). In addition, if students report that they did not have sufficient time to complete the course, this could be a DL program or policy issue if they were trying to complete the course on personal time. Official Army policy is that soldiers should be given duty time to complete required DL courses.

The results of our analysis show that the majority of nongraduations in our sample can be explained by non-DL-related factors, as shown by the two lighter-shaded portions of Figure 2.1.

In 43 percent of cases, the student's nongraduation status was an artifact of recordkeeping. In most of these cases, it was determined that the students had graduated or would graduate soon.[9] Eighty-two percent of respondents in this group had graduated from the DL phase, although their graduation status was not recorded in ATRRS. The remaining 18 percent had taken a course for self-development (CBRN Specialist, Motor Transport Operator BNCOC, CMF 63/94 ANCOC

---

[9] For many students in this category, representing about 10 percent of the nongraduates of DL, we inferred graduation from the fact that they had enrolled in the resident portion of the course.

or AMEDD CCC), for which documentation of graduation or completion in ATRRS is not required.[10]

Another 14 percent of the nongraduations can be explained solely by factors external to the DL program. Six primary external reasons account for nongraduation from DL courses. The two most prominent reasons were "mobilized or deployed" or "changed occupations," each of which accounted for about 30 percent of the cases in this category. Other external reasons for nongraduation (in decreasing order of frequency) include "insufficient time," "left the Army," "emergency situation," and "enrolled in error."

The remaining 43 percent of nongraduations can be tied to the DL program or its policies, either exclusively (31 percent) or in combination with external factors (12 percent).[11] Reasons for noncompletion due to DL program-related factors were grouped into four categories: (1) technical issues, such as insufficient bandwidth or speed or lack of computer access; (2) lack of support for administrative, course content, or technical issues; (3) issues related directly to the course itself, such as course length or content; and (4) insufficient time available to students who worked on required courses on personal time. We discuss these issues further in the next section.

Some students reported both DL and external reasons for not completing the course. The most frequently reported combinations of problems were technical problems with the DL course combined with student mobilization or deployment; technical problems combined with a personal (e.g., student health or family) or work emergency; and a lack of time to complete the course combined with a personal or work

---

[10] The First Sergeant Course was not included for purposes of identifying instances where students enrolled in the course for "self-development." While required for some positions, as an additional skills identifier (ASI) course the First Sergeant Course is not required for promotion. As a result, we feel that "self-development" had a different meaning for that course than for the others. In future versions of the survey, the question regarding the purpose for taking the course should be clarified.

[11] Forty-three percent is a maximum estimate of cases with DL-related reasons for nongraduation. As will be further explained in a section below, this estimate could be slightly lower, but even with a liberal estimate, it would be no lower than 39 percent.

emergency. Our survey did not allow us to determine which reasons might have dominated.

The Main DL-Related Reasons for Nongraduation Were Technical, Support, and Time Issues

While DL-related reasons account for less than half of the nongraduates, improvements in DL policy, approaches, or processes could still potentially make significant improvements in the percentage of enrollees that complete future courses.[12]

Figure 2.2 shows the frequency of DL-related reasons for nongraduation divided into the four categories introduced above. The results show that "technical," "support," and "time" reasons dominated, accounting respectively for 52 percent, 46 percent, and up to 37 percent of the nongraduations.[13] The figure shows that issues related to the

Figure 2.2
Nongraduates: Factors Related to DL Program

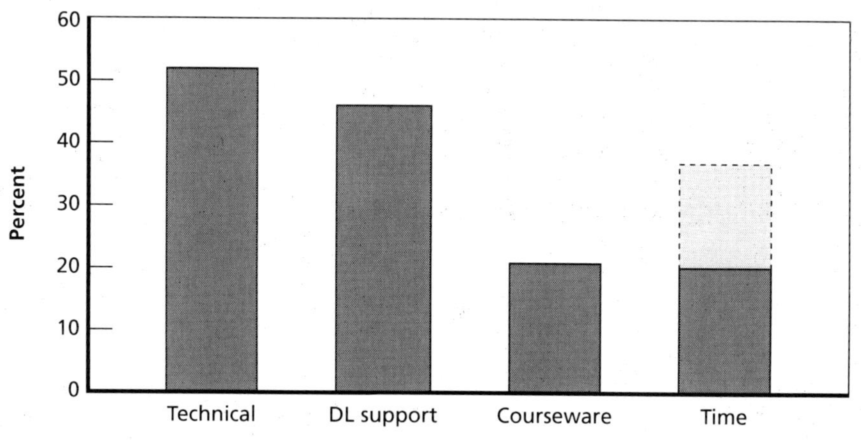

[12] For purposes of relating this analysis to the results in Figure 2.1, we include all participants who cited DL-related issues, regardless of whether external reasons were also cited.

[13] Note that because participants could cite multiple issues, the percentages add to more than 100 percent.

DL courseware itself played a relatively minor role in explaining nongraduation, accounting for only 21 percent of the DL-related reasons.

The right bar in Figure 2.2 is broken into two parts to indicate both a minimum (20 percent) and a maximum estimate (37 percent) of the number of students experiencing DL-related time problems. The minimum value represents students who started the course, cited "did not have enough time" as a reason for noncompletion, and took the course on personal time. The maximum value adds representation of students who indicated they had time problems but did not start the course. The reason for the uncertainty is that students who did not start the course were not asked whether they would have worked on personal or duty time.

Below we further define and explore the specific reasons within each category.

More Than Half of Nongraduates with DL-Related Issues Cited Technical Reasons for Noncompletion

More than half of those citing DL-related reasons for noncompletion (163 participants) cited technical issues to explain that outcome. Nearly three-quarters of these students began the course, and the remainder reported never having started. For the nongraduate survey, we did not collect detailed data on technical problems (see the graduate survey results for more specific measures). But we do know that about one-third of nongraduates with technical issues had trouble getting access to a reliable computer, regardless of whether they started the course or not. Moreover, 22 percent of respondents with technical issues also cited mobilization or deployment as a reason for noncompletion. Among students who did not start the courses, 30 percent had problems getting access to an Internet connection. High-speed Internet access was not a problem for students who started but did not complete courses.

Of the 163 students who reported technical problems, about 25 percent also reported that they could not get answers to their questions about technical issues. These results suggest that many of the remaining students—the 75 percent who had technical problems but did not report a lack of technical support—may not have completed DL courses because they did not know what options they had for technical support or they did not seek technical support when experiencing problems. For example, one respondent simply said, "I become extremely frustrated with computers."

More Than 40 Percent of Nongraduates with DL-Related Issues Cited Problems with DL Support

About 42 percent of those with DL-related reasons for noncompletion (133 participants) cited what we have defined as "support-related" reasons for noncompletion. In 40 percent of these cases, the students reported that they sought help (primarily for technical issues) but did not receive the support they needed.

The remaining 60 percent of cases were students who needed administrative support. They were classified into this category by inference because their responses indicated that simple administrative actions could have addressed their problems. In the majority of these cases, participants were not aware that they had been enrolled in a DL course. These responses point to issues with the existing policy of automatic enrollment in DL courses,[14] suggesting the need for some follow-up.

Students also cited other administrative problems with DL. In 10 cases, respondents stated that they did not receive appropriate materials (presumably CD-ROMs), and a handful confessed that they "forgot" they were enrolled. Finally, 12 participants seemed to lack other basic information about the course, as indicated by responses such as "I was not aware I was supposed to complete the course on my own," "I think I need to re-enroll but I do not know how," and "I do not know if my new unit will support me taking this [1st SGT] course."

Only About One-Fifth of Nongraduates with DL-Related Issues Cited Problems with the Course Itself

About 21 percent of nongraduates citing DL-related reasons for noncompletion (66 participants) said that their noncompletion was due to some aspect of the course itself. Within the category, more than half

---

[14] When students register for the resident phase of a course, they are often automatically enrolled in the DL phase that they are supposed to complete prior to residential attendance. Furthermore, in the case of some courses (e.g., BNCOC or ALC), students can be automatically enrolled in both phases based on their likely promotion.

said that the course was too long. However, the length of the course was only rarely the sole reason for noncompletion. In almost all cases, a response of "too long" was combined with other issues including (in order of frequency) technical problems, insufficient time, lack of student support, or other course-related problems.

Finally, a variety of other course-related factors were mentioned. "Not worth my time" was cited in 30 percent of the course-related cases, and "too difficult" was cited in 20 percent of the cases. Other students commented on a range of issues: "The program taxes the user to wait until the voice finishes reading"; The material is "...not relevant to today's needs"; "The test questions often did not correspond to what was covered in the course"; and "Studying was difficult because course material could not be printed out."

Many Nongraduates Reported Doing DL Coursework on Personal Time

We determined that nongraduates had a DL-related "time" issue when they responded "I found I did not have enough time to work on the course" as a reason for not completing the DL course and also reported that 50 percent or more of their work on the course was completed on personal time (rather than during duty hours). This constitutes a DL-related problem because Army DL policy states that duty time should be provided to take required courses.[15] Sixty-one participants, or about 20 percent of those citing DL-related reasons for noncompletion (see the last bar in Figure 2.2), fit this definition. However, as noted in the discussion of Figure 2.2 above, up to 54 additional cases may also fit.

As a result, our estimate of the number of nongraduates who experienced a DL-related time issue ranges from 20 to 37 percent of the total with DL-related issues (represented by a dashed addition to the bar in Figure 2.2). While we cannot conclusively determine the answers for the missing cases, we note that among nongraduates who did indicate a payment status, 75 percent used predominantly personal

[15] ALARACT (All Army Activities) Message from the Department of the Army (DA) on Army Distributed Learning Policy, February 2006.

time to work on the course. These data suggest that the full extent of the time problem is closer to the upper-bound estimate than to the lower-bound estimate.

Nongraduates Used Substantially More Personal Time for DL Than Did Graduates

To further pursue the effect of the "time issue" on graduation rates, we undertook an additional analysis that compared the results from the surveys of nongraduates and graduates.

Participants in both surveys reported that they used a substantial amount of personal time; i.e., they were not paid for 50 percent or more of the time they spent working on the course (see Figure 2.3). However, nongraduates reported using significantly more personal time than duty time, ($t(105) = -4.51, p < .001$), whereas there were no differences in the use of personal and duty time reported by graduates ($t(303) = -0.52$, *ns*).[16] These results suggest that the amount of personal time used for required DL courses negatively affected graduation rates.

Failure to Complete Courses Is Not Likely to Be Attributable to an Aversion to DL

We also examined the question of whether nongraduates are more likely than graduates to report a dislike for or discomfort with DL. Both the nongraduate and graduate surveys included items measuring students' learning preferences. Items were adapted from existing online surveys of learning preferences (e.g., Online Learning Resource survey (multiple universities); Peterson's Distance Learning Assessment; Participant Perception Indicator). Eight items were measured on 5-point

---

[16] Here, the t-statistic was used to test the null hypothesis that the difference between responses (personal time and duty hours) within each group has a mean value of zero. Obtaining small p-values (typically $< .05$) allows one to reject the null hypothesis. The notation "*ns*" refers to results that are not statistically significant.

Repeated-measures analysis of variance shows a significant interaction between graduate status and payment status, $F(1,408) = 11.63, p < .001$. The F-statistic indicates whether the variances between the means of two or more populations are significantly different from each other. In this analysis, the result shows that the difference between duty hours and personal time is significantly greater for nongraduates than for graduates.

Figure 2.3
Payment Status (Nongraduates and Graduates)

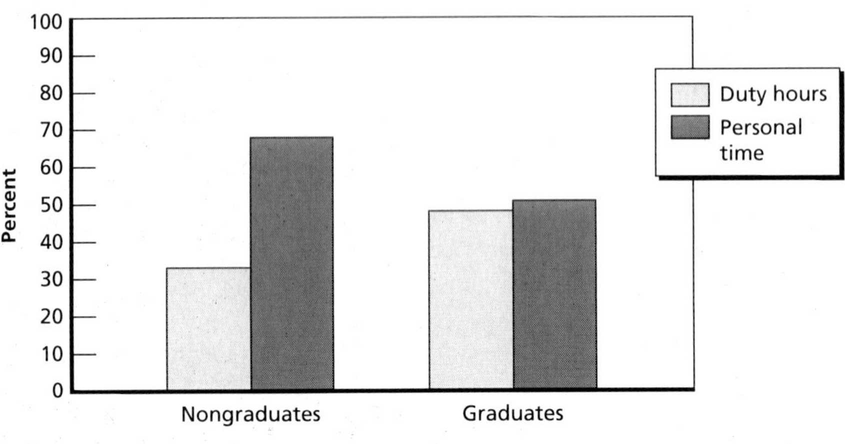

RAND *MG1072-2.3*

scales, ranging from strongly disagree to strongly agree (Question 14 in Appendix A and Question 49 in Appendix C). Factor analysis of responses across both surveys revealed two distinct factors, one that we call "DL orientation" reflecting a preference for independent, self-paced, computer-based learning, and one that we call "classroom orientation" reflecting a preference for classroom-based learning that is guided by an instructor and involves interaction with other students.[17] All students had scores for both DL and classroom orientation.

Figure 2.4 shows average ratings of DL and classroom orientations for nongraduates and graduates. On average, participants in both groups preferred classroom learning (M = 3.83, SD = .63) over DL

[17] Principal axis factor analysis with a varimax rotation was conducted using responses from both surveys. Results revealed two factors that account for 62 percent of the variance in responses. Factor loadings ranged from .75 to .86 for DL orientation and .62 to .83 for classroom orientation. Coefficient alphas for the two scales were .82 and .71, respectively. The factors were slightly negatively correlated with each other, $r = -0.09$, $p < .05$.

The survey included a measure of comfort with technology (Agarwal and Prasad, 1998). Scores on this scale were highly correlated with DL orientation and are not analyzed further.

Figure 2.4
Ratings of Learning Preferences (Nongraduates and Graduates)

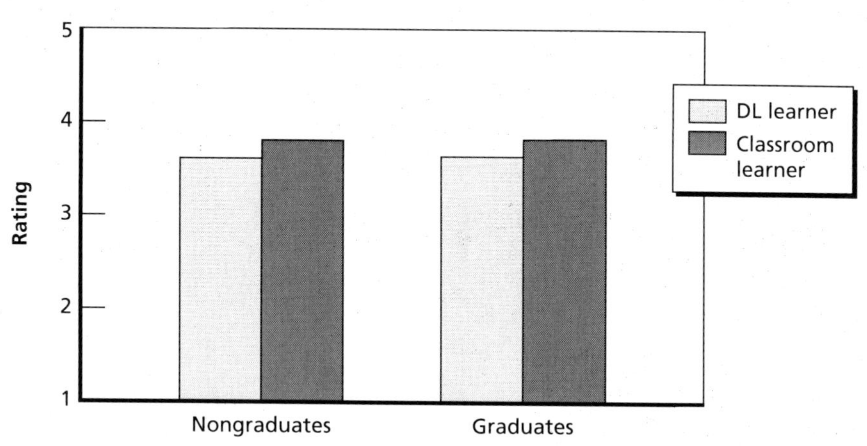

RAND *MG1072-2.4*

(M = 3.63, SD = .77).[18] Although differences between classroom and DL orientations were statistically significant for both nongraduates and graduates, we note that the absolute average rating of DL orientation is generally high. Moreover, there were no differences between nongraduates and graduates in relative ratings of classroom and DL orientation, $F(1,744) < 1$. These results suggest that failure to complete courses cannot be attributed to an aversion to DL.

Now that we have assessed reasons for nongraduation, we present results from analyses of the graduate survey examining students' reactions to DL.

Students Are Moderately Satisfied with Their Experience in DL

Figure 2.5 shows average ratings of satisfaction for multiple aspects of students' experiences in DL courses. On all five measures, the average rating was greater than 3 on a 5-point scale, suggesting that students were moderately satisfied with all aspects of the course that we assessed. Below we provided more detailed results for these measures.

---

[18] M = mean or average score. SD = standard deviation. The SD indicates how much variation there is around the mean score.

Figure 2.5
Graduates' Satisfaction with Different Aspects of DL

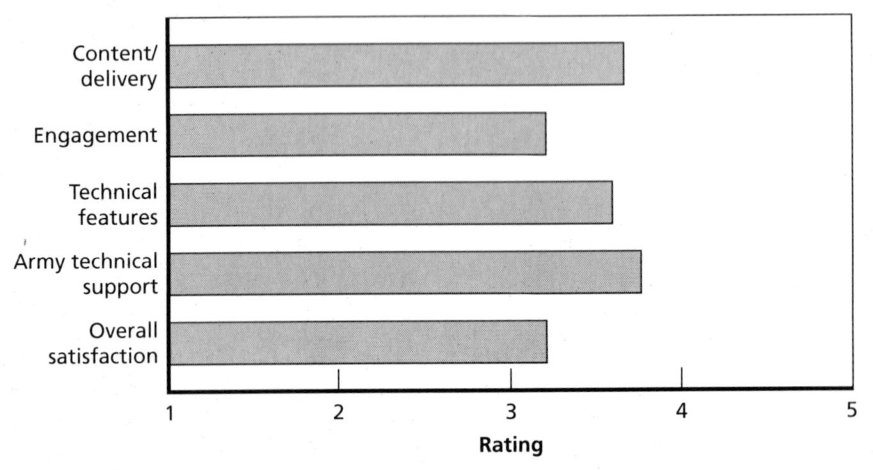

RAND *MG1072-2.5*

Graduates Were Generally Satisfied with Course Content and Delivery, But Ratings of Learner Engagement Suggest Some Need for Improvement

The graduate survey included 12 items about the quality of instruction, which were grouped into two scales:[19]

- The first scale, content/delivery, consists of eight items measuring perceptions of the clarity of explanations of concepts and demonstrations of procedures, number and quality of practical exercises, value of feedback, applicability of course to the job, and value of the course in preparing students for subsequent training.
- The second scale, engagement, consists of four items reflecting the extent to which the course was interesting and held the learner's attention.

[19] Items with low item-total correlations for these and other scales were dropped. The remaining items form internally consistent scales, with Cronbach's alpha ranging from .74 to .95. Cronbach's alpha is a measure of reliability or consistency and reflects the extent to which items measure a similar construct.

The average rating for quality of instruction was 3.67 and the average rating for learner engagement was 3.21 on a 5-point scale (*SDs* = .63 and .77, respectively).

Most Graduates Thought Course Length and Level of Difficulty Were About Right

Two of the learner engagement items on the graduate survey had follow-up questions to better characterize students' experiences in DL. Students who disagreed with the statement "The length of the course is about right" were then asked if the course was too long or too short; likewise, students who were not satisfied with the course's level of difficulty were asked if the course was too easy or too hard. As shown in Figure 2.6, about half of the students thought that the length of the course was "about right," and close to 70 percent of the students thought the level of difficulty was appropriate.

A number of graduates provided specific comments about the quality of instruction or had suggestions for improvement. Ten commented that the course should be moved back to a classroom setting, eleven wrote that many of the modules were too long and should be

Figure 2.6
Graduates' Ratings of Course Length and Difficulty

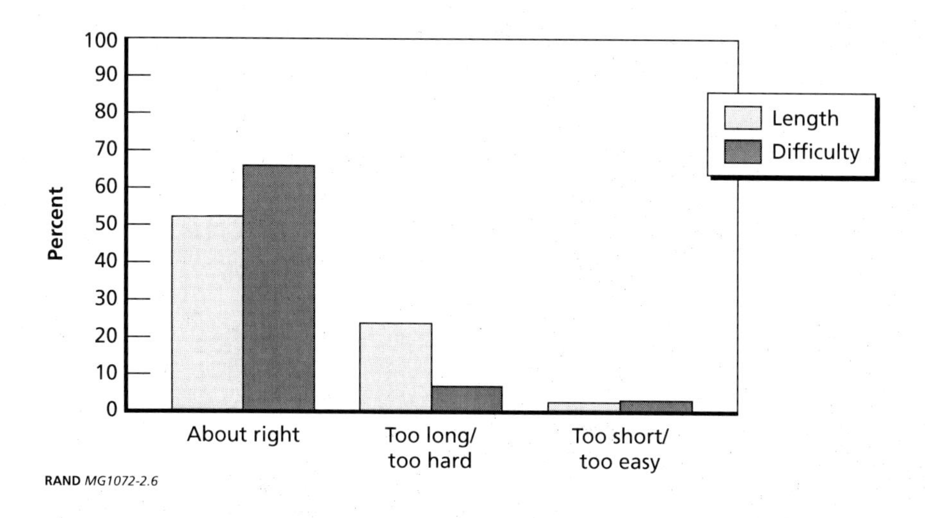

broken down into smaller sections, and ten felt that the course material was too difficult. Eleven students commented that information in the course, ranging from grammar to course content, was incorrect or that material was outdated.

The Vast Majority of Graduates Felt There Was Too Little Interaction with Instructors and Peers

The graduate survey also included questions about the degree of and satisfaction with opportunities for interaction with instructors and other students. Sixty-three participants (15 percent) reported that they interacted with course instructors about course content, and 119 (29 percent) reported interacting with other students. These respondents, who were primarily from the Armor and Battle Staff NCO courses, were generally satisfied with the level of interactions. However, the preponderance of graduates across all courses reported that there was too little interaction with instructors and peers (see Figure 2.7). These results are consistent with some comments from nongraduates, who reported that the courses were too long to undertake without instructor support.

Figure 2.7
Graduates' Satisfaction with Interaction

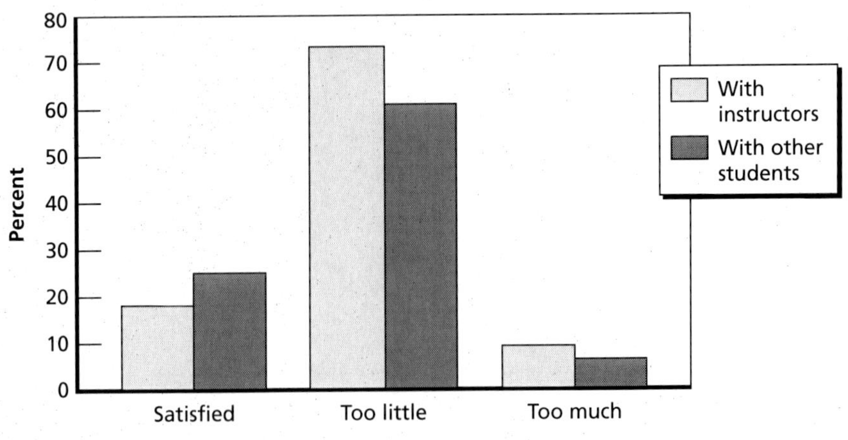

RAND *MG1072-2.7*

Graduates Were Moderately Satisfied with Technical Features of Courseware

Graduates were asked to rate their overall satisfaction with the technical features of the course on a 5-point scale. Students were moderately satisfied with the technical characteristics of the courseware overall; Figure 2.5 above shows that the average satisfaction rating was 3.55 (*SD* = .88). There were no differences in ratings between students who took courses online or by CD-ROM. However, 63 percent of respondents reported some technical difficulties (as explained below), suggesting some need for improvement. More detail about the nature of the technical problems is provided below.

Predominant Technical Problems for Graduates Include Bandwidth and Access to Courseware

Graduates were asked if they experienced any of 13 technical problems while taking the course. Problems were grouped into one of four categories: bandwidth/speed, access to the course, difficulty with navigation, and production quality. Figure 2.8 shows the frequency with which students experienced these problems for students who took the course primarily (75 percent or more of the course) online using high-speed Internet access or who took the majority of the course on CD-ROM.[20] (Twelve percent of all respondents fell into the CD-ROM category, and only the Battle Staff NCO course had a sufficient number of students who used CD-ROMs to report meaningful results.)

Thirty-seven percent of the students who took the course online had no technical difficulties with the courseware. For students who had problems, the most common issues pertained to bandwidth or speed (e.g., delays in pages loading; difficulty playing audio or video files), followed by access to courseware (e.g., difficulty launching the course or receiving CD-ROMs). Fewer students reported problems navigating through courseware. Production quality (e.g., difficulty reading the text, excessively slow or fast narration) was the least problematic cat-

---

[20] We excluded responses from a small number of students who reported using dial-up connections or a substantial mix of media (e.g., 50 percent high-speed Internet and 50 percent CD-ROM).

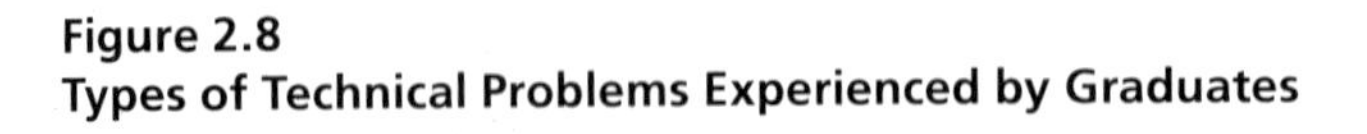

Figure 2.8
Types of Technical Problems Experienced by Graduates

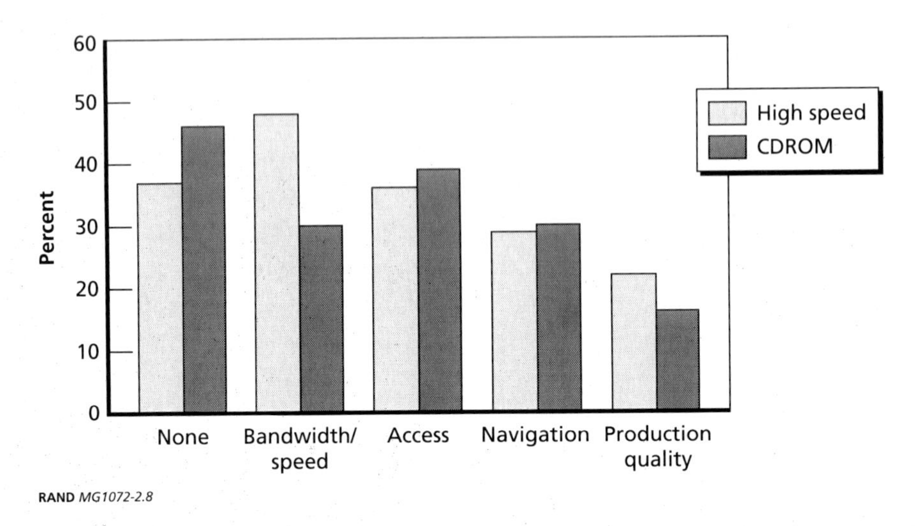

RAND *MG1072-2.8*

egory of technical difficulties. Students who had technical problems with online courseware reported anywhere from 1 to 11 problems, with an average of 3.41 problems (*SD* = 2.42). Students tended to report combinations of problems involving difficulties related to bandwidth and navigation.

Forty-six percent of graduates in the Battle Staff NCO course who took the course primarily by CD-ROM reported no technical problems. Students who had technical issues reported anywhere from 1 to 7 problems, with an average of 2.93 problems (*SD* = 2.02). As with the online students, some issues were cited much more frequently than others. The most frequently reported problem was difficulty returning to the place where the student left off after exiting and returning to the course. The second most common problem was delays in pages loading. There were no systematic patterns in combinations of problems among students taking the course on CD-ROM.

Some graduates also cited other technical difficulties that were not mentioned in the survey questions. Common problems were reported by students from two courses: 15 AMEDD CCC students reported difficulties in printing their course completion certificate, and 6 students from the Cavalry Scout course reported that the navigation module was difficult to understand and complete due to slow Internet connections and difficulty zooming in on the map. Other responses across all courses included difficulty loading the courseware when deployed and requests for improved animation and graphics. Several graduates recommended providing a low-bandwidth version of the course (i.e., CD-ROM or DVD) for use with slower Internet connections when deployed. Related suggestions were to break longer lessons into smaller sections and eliminate some of the photos to reduce delays in loading pages over slow Internet connections.

Graduates Were Generally Satisfied with Technical Support, Especially Help Desks

Graduates were asked if they needed technical support, and if so, what type of support they used and their level of satisfaction with it (see Table 2.3). Seventy-four graduates (17 percent), primarily from the AMEDD CCC and Battle Staff NCO course, reported using technical support, and about one-half of these students reported using more than one source of support. The most frequently used sources included the Army Training help desk and proponent schools' help desks. As shown in Table 2.5, students were generally satisfied with each of the sources of support, although experience with help desks received higher ratings than did experience with other sources. As shown in Figure 2.5, the average rating of overall satisfaction with Army technical support (excluding "other" sources) for these students was 3.77 (*SD* = .90), indicating that students were moderately satisfied with the technical support that they received.

Table 2.5
Types and Quality of Technical Support Used by Graduates

Army Training Help Desk		Proponent School Help Desk		Contacted Instructor		Other Source	
n used	Quality	*n* used	Quality	*n* used	Quality	*n* used	Quality
53	3.75 (.94)	41	3.78 (.73)	21	3.38 (1.17)	15	3.07 (.80)

Most Graduates Did Not Recall Using Supporting Materials

Several questions were included to assess the value of supplementary instructional resources, such as field manuals and glossaries, as well as other support for course content. Graduates were asked if the course had field manuals or glossaries, if these materials were useful, and if they were able to get a satisfactory answer about content from either the supporting materials or an instructor. The majority of students (53 percent) reported that their courses did not have supporting materials or that they didn't remember if supporting materials were available—in spite of the fact that most courses had such resources. On average, students who reported that these materials were available were satisfied with the quality of support (M = 3.97, SD = .87). On average, students who reported needing other support for content (61 percent) were moderately satisfied with the support provided (M = 3.39, SD = 1.00).

Several graduates provided comments related to supporting materials. Two of these students noted that while the course contained references to supporting materials, none were provided or links to other resources on the Internet were no longer active. A number of students recommended providing access to the course materials on CD-ROM or hardcopy for later reference.

Overall Satisfaction of Graduates Was Most Strongly Associated with Level of Learner Engagement

Overall satisfaction was measured with four items, e.g., "Overall, I was satisfied with the DL phase of this course" and "I look forward to taking another DL course." The average rating, as shown in Figure 2.5, was 3.21 (SD = .77). Overall satisfaction was strongly associated with other measures reported in Figure 2.5. Overall satisfaction was most strongly correlated with learner engagement (r = .79), followed by ratings of quality of instruction (r = .65), technical features of courseware (r = .47), and Army technical support (r = .45). All correlations were statistically significant at $p < .001$.

Some graduates provided written comments about their overall impressions of the course. In general, these students were evenly split in terms of positive and negative attitudes. Twenty students commented that the course was not worthwhile, was a waste of time, or that the material was not useful, whereas 17 commented that the course was worthwhile, relevant, and useful. Some examples are:

- "The course was generally redundant, did not give me any new practical knowledge."
- "I felt the DL course was very through [sic] and prepared me for the resident phase as well if not better than anything else I could have done. I felt very confident going into the resident phase."

Graduates with Stronger Preferences for DL or Classroom Learning Reported Better Experiences in DL

We examined the extent to which students' preferences for DL and classroom learning influenced their satisfaction with DL courseware. This analysis was conducted to determine whether there are individual trainee characteristics that affect reactions to DL training. A positive association between DL orientation and satisfaction and/or a negative association between classroom orientation and satisfaction, for example, might be used to target different types of courses to different students or to design interventions to influence student learning preferences.

We compared ratings of overall satisfaction for students with high and low scores on the DL orientation scale. As shown in Figure 2.9, students with stronger preferences for DL were much more satisfied with the course. However, a comparison of students with high and low scores on the classroom orientation scale shows a virtually identical pattern. These results suggest that students who are more oriented toward learning—whether classroom or DL—were more satisfied with their experience in DL. Thus, like the findings regarding the association of learning preferences and graduation status described earlier, these results suggest that interventions designed to influence students' receptivity to DL would not be worthwhile. However, research evidence suggests that it would be beneficial to use interventions that encourage students' *mastery orientation*, or goals that emphasize learning for self-improvement, skill development, and long-term competence (e.g., Dweck, 1986), as opposed to an emphasis on learning in order to perform well in training (Mesmer-Magnus and Viswesvaran, 2007). Emphasizing learning goals is valuable in any delivery mode, but it

Figure 2.9
Overall Satisfaction Ratings for Graduates by Orientation: DL and Classroom

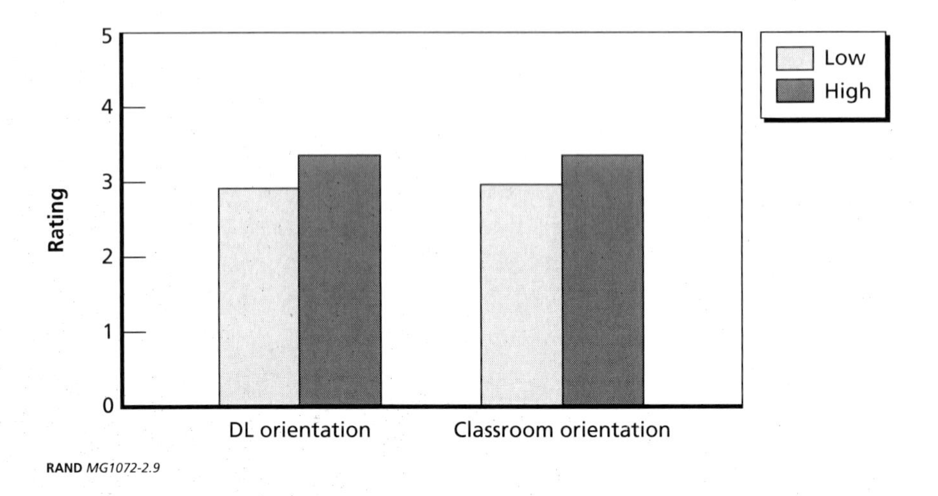

RAND *MG1072-2.9*

appears to be especially important in Web-based and blended-learning environments where students have substantial control over when and how learning takes place (Klein, Noe, and Wang, 2006).

Conclusions and Recommendations for Improvement in DL Courseware and Supporting Systems

Overall, responses to both the nongraduate and the graduate surveys indicate relatively favorable DL outcomes. At the same time, survey findings point to several areas for improvement in DL courseware or supporting systems.[21]

---

[21] We note that the primary intent of this effort was to develop and pilot-test surveys of students' experiences in DL and to demonstrate the kinds of evaluation information that these surveys can provide. Given that each survey was implemented in a small number of courses, recommendations for changes may not be generalizable across TADLP. Thus, we describe our recommendations as "suggestive" rather than "definitive."

The Army Needs Better Capture of Information Regarding Student Status in Courses

The nongraduate survey shows that actual graduation rates are higher than those derived from ATRRS. We make the following recommendations for improving the capture of information about student status in DL courses. These changes in record keeping will produce more accurate and favorable statistics about DL usage.

- First, we recommend adding a field to ATRRS to document the student's purpose in enrolling in a course (i.e., requirement, self-development, refresher/reachback), which would be completed by the student when enrolling for the course.
- Second, the Army should encourage updating of the ATRRS course graduation field for DL courses or phases of courses. Ideally, course completion would be auto-populated from the learning management system (LMS) used to deliver the DL course, but a current lack of interoperability among Army training information systems may require manual input of this information. Another option is to require that ATRRS have a valid graduation status for the DL phase in order for students to graduate from the resident phase of the course.
- Third, the problem of students being enrolled in courses without their knowledge or forgetting that they have enrolled could be eliminated or reduced by having the system send an automated notice to students who have enrolled but not "touched" a course within a specified time period or by sending reminder notices if ATRRS does not show that the student has graduated within a set amount of time (e.g., after 6 months). Presumably, populating a field regarding students' status in DL courses would require interoperability between the LMS and ATRRS. Automated notices could also be made more effective by ensuring that the student's Army Knowledge Online (AKO) email address is correct, preferably by auto-populating or auto-verifying the AKO address. Currently, ATRRS verifies that students enter a us.army.mil address; however, if the information before the "@" sign is incorrect, then the message will not be delivered. In addition, our

results showed that some students were enrolled by someone else, so automated notices from ATRRS were sent to the enroller, not the enrollee.

The Army Should Enforce the Policy of Paying Soldiers for Required Training

Both surveys found that a substantial amount of DL training—even for required DL—was done on personal time rather than on duty time (as specified by DA policy). This pattern was especially pronounced for nongraduates. One option for changing the pattern is the current effort to create an "EDY" or educational duty status that students can use while working on required courses. Another option observed in two courses in our graduate survey was used by the Army National Guard (ARNG), which allowed students to come to the schoolhouse one week before the resident course to complete the DL prerequisite on paid time. This approach appeared to increase the likelihood that students completed the DL phase and did so prior to the start of the resident phase. Other options may also be needed.

In addition to complying with Army policy regarding payment for time spent on training, the ARNG approach has other benefits. First, compared to students who complete the DL phase several months in advance, these students were more likely to retain knowledge of what they had learned, enhancing their readiness for resident training (see Chapter Three). Second, students' comments in the graduate survey indicated that they benefited from the opportunity to interact with instructors and peers about DL course content at the schoolhouse. Thus, these students appeared to have an experience that loosely resembles blended learning, which mirrors one of the strategic directions for training in the Army Learning Concept (TRADOC, 2011). Although the ARNG's approach in these courses sacrifices the anytime/anywhere benefit of DL, it may ultimately be more beneficial than having students complete DL over longer periods of time and without instructor and peer support. As described earlier, research findings show that the quality of human interaction during training is strongly and positively related to student reactions (Sitzmann et al., 2008). In addition, Sitzmann, Kraiger, Stewart, and Wisher (2006) found that students acquired more declarative and procedural knowledge in blended learning than in Web-based or classroom instruction. Therefore, the ARNG approach may enhance both learning from and reactions to training.

If enforcement of the policy to provide duty hours for DL is deemed unattainable, then TADLP should consider implementing policies that change its DL model in order to avoid the issue; for example, a specific type of blended-learning option calls for students to complete DL content while in residence. The Navy reports that converting a portion of resident A-School courses to a computerized self-paced format resulted in 10–30 percent reductions in time to train (Carey et al., 2007).

The Army Should Seek to Improve Learner Engagement and Specific Technical Features in DL Courseware

The graduate survey showed that students were moderately satisfied with multiple aspects of their experience in DL, although with some variation by course. However, ratings of some features of DL courses also point to recommendations for the design of future courseware. Designing DL to provide for more opportunities for interaction with instructors and peers is likely to increase student engagement in IMI and may also provide students with connections to people who can provide help with technical or substantive issues. The Army's current move toward blended learning may achieve these goals. Student engagement may also be fostered to the extent that the course provides sufficient numbers of examples from the job or mission environment and opportunities for practice (Straus et al., 2009). Engagement may also be enhanced by decreasing course length, for especially long courses (e.g., by allowing qualified students to test out of modules).

The large percentage of students citing technical problems suggests some need for improvement in particular areas. Most important is addressing the bandwidth problem, followed by the problem with courseware access (e.g., difficulty launching). Providing low-bandwidth versions of courses or CD-ROMs for students in deployed locations or in other constrained settings is one possible short-term solution to the first problem; in fact, some students commented that they would have preferred to have the course on CD-ROM either to circumvent bandwidth problems or to use as a resource (e.g., for reachback). While there might be some concern that lower bandwidth courses will sacrifice features of IMI that enhance student interest (such as videos), engagement need not come from complex or data-intensive media but can come from better pedagogy.

Future Administration of Surveys

Our findings indicate that both the graduate and nongraduate surveys are appropriate for ongoing use at the program level. The scales are psychometrically sound in terms of providing reliable measures, and results show a reasonable degree of variation in responses (rather than showing uniformly high or low ratings, for example). From an administrative standpoint, the surveys are not burdensome to complete. The average completion times for the nongraduate survey ranged from 6.5 to 8.7 minutes (depending on students' course status) (*SDs* = 3.46 and 4.31). The average time to complete the graduate survey was 12.9 minutes (*SD* = 6.15).[22] In addition to answering the objective (i.e., closed-ended) questions on the graduate survey, 229 students provided substantive written responses to at least one open-ended question, and numerous students answered more than one open-ended question. In addition, we shortened the graduate survey by eliminating questions that were included exclusively for research purposes and items that do not appear to measure what they were intended to measure (see Appendixes F and G for the revised surveys). Thus, the revised graduate survey will take less time for students to complete.

Both surveys also are straightforward to interpret and score. Scoring instructions for the revised surveys are presented in Appendix H. In addition, the surveys can be adapted to address specific goals or topics of interest to TADLP or individual proponent schools.

[22] These results exclude participants with extremely long completion times (greater than 45 minutes), as we suspect that these participants may have been engaged in other activities while working on the survey, and extremely short completion times (less than 5 minutes), which suggest that participants clicked through the survey or did not answer all the questions.

Some considerations regarding use of the surveys include timing of administration and recruitment. Some of the courses in the pilot implementation of the graduate survey were convenience samples of students who were recruited some time after they may have finished their DL phase (e.g., on the first day of follow-on resident training). However, the graduate survey is intended to be administered when students complete the DL training.

We make some additional recommendations to improve the use and value of the surveys:

- First, all schools should use the items in Appendixes F and G to provide TADLP with a common set of indicators, with the option for schools to add questions to address local interests. Providing opportunities to customize the surveys may help enhance buy-in from the proponent schools.
- Army Training Support Center (ATSC) should provide a common platform and software application to enable the schools to design, administer, and score the surveys online. The Army's Battle Command Knowledge System (BCKS) has such a capability and could potentially provide it to ATSC.
- Use of the graduate survey should be part of routine quality improvement efforts. We recommend providing a link to the survey on the course LMS and asking students to complete the survey before printing their certificates of completion. At the same time, the Army may wish to consider using an alternating schedule of survey administration or to make the survey optional in some classes of a course to avoid "over-surveying" students and to minimize perfunctory responses.
- Nongraduates should be contacted via email to complete the nongraduate survey one year after enrollment—the time that students have to complete the course. Again, to avoid over-surveying, the survey might be administered to a random sample of nongraduates. A modified version of the nongraduate survey might be administered sooner, such as at the six-month point following enrollment, with one of the goals being the identification of students who might need help.

- The nongraduate survey found that many students did not ask for help when they needed it. Student support might be improved by giving students the opportunity to provide immediate feedback or submit questions via email or online chat (e.g., to the Army Help Desk) as they encounter a particular issue with a course.

CHAPTER THREE

Knowledge Retention of DL Material in the Phased Approach to Training

"Learning" refers to the acquisition of new knowledge, skills, and attitudes and is a (or perhaps the) primary objective of training. Changes in knowledge, skills, and attitudes can foster improvement in individual job performance and organizational effectiveness as well as contribute to societal goals more generally (Goldstein and Ford, 2002).

Comprehensive evaluation of training should include measures of three types of learning outcomes identified by Kraiger, Ford, and Salas (1993): cognitive, skill-based, and affective.[1] These outcomes correspond to cognitive learning, training performance, and post-training attitudes shown in Figure 1.1:

- Cognitive learning includes verbal knowledge (declarative knowledge or knowledge about "what," procedural knowledge or knowledge about "how," and knowledge about task performance strategies); knowledge organization such as mental models; and cognitive strategies, which reflect the speed with which knowledge can be accessed or applied.
- Skill-based learning includes skill acquisition, which reflects a transition from declarative to procedural knowledge; skill compilation, which is evidenced by faster and more accurate performance and integration of separate steps into single acts; and automatic-

[1] Kraiger, Ford, and Salas (1993) draw on taxonomies created by Bloom (Bloom, Englehart, Furst, Hill, and Krathwohl, 1956) and Gagne (1984).

ity, in which an individual can perform a task without conscious monitoring and can carry out other tasks simultaneously.
- Affective learning refers to attitudes and motivational states that constitute objectives of the training course or program. For example, a course may endeavor to foster positive attitudes about working in demographically diverse teams or about complying with organizational policies for computer security. Self-efficacy, which refers to one's perceived capabilities to perform a specific task, is an example of an affective motivational outcome; changes in self-efficacy can be an indicator of training success. Kraiger, Ford, and Salas (1993) discuss measurement implications for all three types of learning.

Our focus in this chapter is on cognitive outcomes. Cognitive learning in general, and declarative knowledge in particular, are necessary first steps in acquiring higher-order knowledge and skills (e.g., Ackerman, 1987) and influence readiness for subsequent training and performance on the job. Therefore, understanding the degree to which students acquire declarative knowledge and retain it is a first and important step in documenting the impact of DL and identifying areas for improvement in training content and delivery.

In this chapter we discuss the results of a pilot study assessing delayed procedural knowledge, or knowledge retention, from IMI, and its association with other variables in the Alvarez, Salas, and Garofano (2004) model (see Figure 3.1, adapted from Figure 1.1). Most Army IMI focuses largely on declarative knowledge (Straus et al., 2009). We address cognitive learning from training, its association with soldier readiness for subsequent training (transfer performance), and characteristics that influence these outcomes (see the variables that appear in boldface type in Figure 3.1). We tested the relationships among these variables using knowledge retention tests in two courses to illustrate and explore use of the model and the challenges involved with its implementation.

Figure 3.1 is a conceptual illustration of the relationships we would ideally want to study to examine the relationship between learning and readiness. The framework posits that students who perform better in

Figure 3.1
Subsection of Alvarez, Salas, and Garofano (2004) Integrated Model Used to Guide Studies of Knowledge Retention

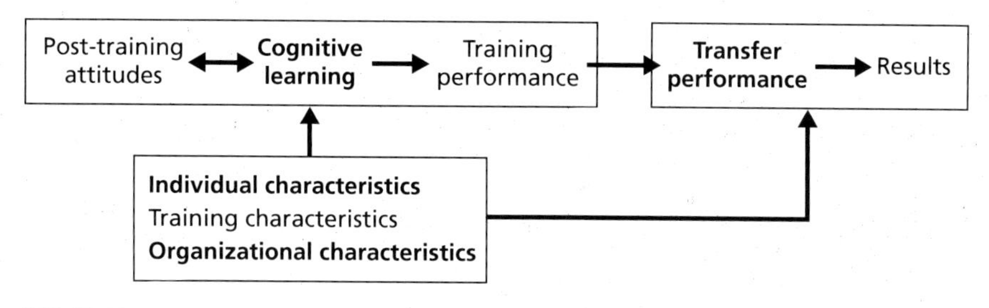

RAND *MG1072-3.1*

training (in terms of changes in or acquisition of knowledge, skills, and attitudes) will show higher transfer performance, or individual readiness. Transfer performance typically refers to job performance, but can also be construed as readiness for subsequent training.

Individual characteristics that affect these outcomes include factors such as cognitive ability and relevant job experience. Examples of relevant training characteristics include course content, instructional design, and delivery modes. Relevant organizational characteristics include the circumstances surrounding training, such as the length of time spent in training, the location where training occurs, and the amount of time or lag between DL and resident training. Some of these circumstances are determined by training policy—for example, as described in Chapter Two, ARNG policy influences payment status, location, and time spent on training by allowing students in some courses to complete DL at the schoolhouse one week prior to resident training. Circumstances like location can affect learning if, for example, the location provides access to instructors and peers who provide support.

In the Army's phased approach to training, we expect that a shorter time lag between completion of DL training and the start of resident training may enhance knowledge retention from DL and therefore enhance preparation for the resident phase. On the other hand, circumstances can negatively influence learning outcomes if, for example, a student has only limited time available for training and has to "cram."

Challenges in Examining the Relationship Between Learning and Soldier Readiness

The Army has undertaken its own efforts to measure the effects of learning on performance. In particular, the Army currently uses the AUTOGEN program in an effort to assess the contribution of all training (not just DL training) to job performance.[2] AUTOGEN consists of an automated system that enables proponent schools to develop and conduct online surveys. Surveys are completed by unit supervisors, who are asked to evaluate how effectively a group of students can perform critical tasks (terminal learning objectives) from specific training courses. Students also complete these surveys regarding perceptions of their own performance.

However, discussions with training staff indicate that supervisor response rates to AUTOGEN surveys are low. Furthermore, AUTOGEN does not distinguish results for DL and resident training and therefore cannot provide information about the contribution of DL for courses that use multiple delivery modes. AUTOGEN also does not directly link performance in training with behavior on the job; thus, there is no way to determine whether student performance in units is due to training or other factors.

Even if individualized data on training performance could be brought into the AUTOGEN analysis, there are other challenges to establishing the link between performance in training and behavior on the job. One obstacle is that some soldiers do not perform the job for which they were trained, thereby limiting the potential to collect relevant performance data for trainees. Assessing the link between training and behavior is also difficult due to the typical lack of variability in performance in Army training. Scores on written tests generally range

---

[2] AUTOGEN (Automated Survey Generator) is licensed by the U.S. Army Research Institute (ARI).

from 70 to 100, and performance on skills tests are rated "Go" or "No go," with most students receiving ratings of "Go." A lack of variability in scores, or restriction of range, limits the potential to observe associations between scores in training and other outcomes.

Assessment of Knowledge Retention from DL

Demonstrating the association of DL with job performance was beyond the scope of our present research effort. Instead, we tested other ways to assess the value of DL. Specifically, we sought to study the impact of DL by examining immediate cognitive learning, delayed cognitive learning (Alliger et al., 1997) (i.e., knowledge retention) and the associations of learning and knowledge retention with readiness for (i.e., performance in) follow-on resident training. As shown in Figure 3.2, we propose that cognitive learning in DL training will be positively associated with knowledge retention, which in turn will influence performance in follow-on resident training. While the model would apply equally well to "skills" as to "knowledge," measuring declarative knowledge is especially appropriate early in training. In addition,

Figure 3.2
Model Tested in Assessments of Knowledge Retention

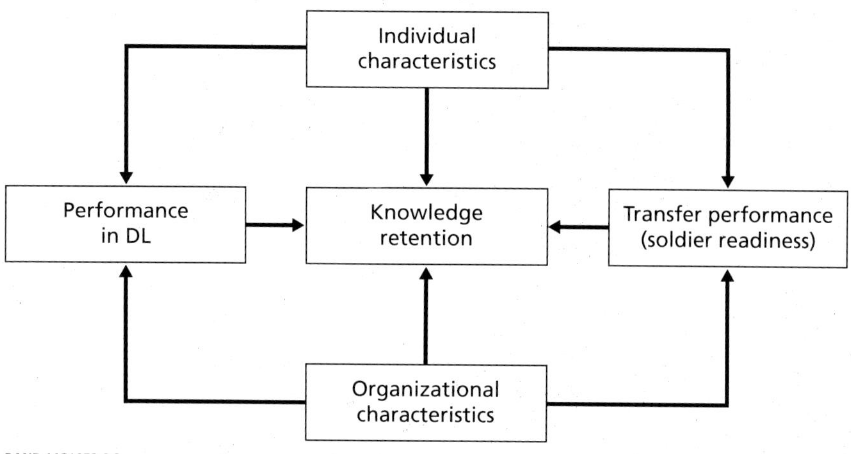

by focusing on courses that use written knowledge retention tests, we were able to obtain measures of training performance with sufficient variability for analysis.

In addition, we posit that both individual characteristics and organizational characteristics affect all three outcomes (see Figure 3.2). However, because we examine individual courses, we do not include measures of training characteristics in the assessments because these would require comparing multiple courses that vary in factors such as delivery modes.

Our assessments of knowledge retention sought to answer the following questions:

1. What is the level of knowledge retention following the DL phase?
2. What is the association of performance in the DL phase and knowledge retention?
3. How does the lag following the DL phase affect knowledge retention?
4. Does DL affect preparedness for subsequent training?
5. How are other training policy and student characteristics associated with learning and individual soldier readiness?

Method

We developed knowledge retention tests for two high-priority courses that use DL in the phased approach to training: Ordnance Mechanical Maintenance Basic Knowledge and Skills Course (63/91 A, B, D, H, M [R]) and the Battle Staff NCO course.

Ordnance Mechanical Maintenance Course. The Ordnance 63/91 DL course teaches basic mechanical maintenance skills and serves as a prerequisite for five reclassification courses taught in residence. At the time of the study, the course was using a written assessment of the DL phase administered on the first day of resident training, i.e., a test of DL knowledge retention. The test consisted of 50 multiple choice items as well as questions measuring training circumstances and student characteristics. Items used in our analysis included civilian mechanical maintenance experience,[3] and experience in current (i.e., reclassification) military occupational specialty (MOS).[4]

The Ordnance school provided RAND Arroyo Center with test scores from 272 students. Students' scores on four tests from the DL phase of the course and DL completion dates were provided by ATSC. These data also were used to create an estimate of time spent on the DL course, calculated as the number of days between the first and last course tests. The start date for resident training was obtained from ATRRS, and student demographic characteristics and Armed Forces Qualification Test (AFQT) scores were obtained from the Total Army Personnel Database (TAPDB) and the Reserve Components Common Personnel Data System (RCCPDS).

Challenges in obtaining needed data from different Army information systems led to a reduction in the scope of our analysis. Approximately 50 students could not be identified in ATRRS or the personnel databases with the identifying information available to RAND (name and last four digits of the social security number, SSN), even after manually reviewing 80 recent training rosters from ATRRS to find possible matches.[5] In addition, data from 54 students could not be used because these students did not complete the DL phase prior to resident training. Therefore, analyses using some student characteristics and training policy/circumstances variables are based on approximately 130 students.

Battle Staff NCO Course. The Battle Staff Course trains NCOs how to operate as part of a battalion or higher staff and consists of both a DL (IMI) phase and a resident phase taught at Fort Bliss or at several regional training sites via video teletraining (VTT). In contrast to the Ordnance course, the Battle Staff course had not been using a

---

[3] Options included: no experience; shade tree mechanic; light vehicle servicing (Jiffy Lube, Tuffy, Sears, Walmart, etc.); and major dealership service department or industry (Ford, CAT, Cummings, etc.).

[4] Options included: I have not been working in the new MOS; less than six months; seven months to one year; one year to two years; over two years.

[5] We did not request full SSNs for privacy and security reasons.

test of knowledge retention. Therefore, we designed an online knowledge retention test in collaboration with the course manager. The test was based on checks on learning and items from the end-of-course test from the DL phase. The knowledge retention test consisted of 50 multiple choice items as well as questions about the student's training circumstances.

We pre-tested the test in a class of 19 ARNG students. Thirteen students completed the test (68 percent) within the first few days of the start of resident training. The average score on the test was 53 percent correct (*SD* = 13), suggesting either that the test was difficult, knowledge retention was poor, and/or students were not motivated to perform well. Based on the responses and input from the course manager, some test items or response options were revised; however, the course manager felt that many students were incorrectly answering questions on concepts that they should have known from the DL phase.

We recruited 98 students from three VTT classes to take the revised knowledge retention test. Time was provided for students to complete the test at the beginning of training. 61 students (62 percent) participated in the study. Responses and system logs indicated that 20 students either did not complete the test or rushed through it (i.e., completed the test in less than 7 minutes), so students were asked to retake the test if they had "clicked" through it. Ten of these students completed the test, and their first attempts were deleted. Subsequent analyses were conducted on the remaining 51 cases. The average score on this test was identical to that of the pre-test group (*M* = 53 percent, *SD* = 12).

The course manager provided student performance data from the end of the resident phase. We obtained the data on student characteristics and training circumstances using the same data sources described above for the Ordnance course. However, 16 students had not completed the DL phase prior to resident training, so they did not have data for many of the variables in the model. This left only 35 students for analysis, which is too small a sample to answer our research questions with confidence. Therefore, we do not present findings from this course in our results. We do, however, discuss a number of lessons learned about measuring the impact of DL training on learning outcomes.

Findings

Data from the Ordnance course were used to assess the level of knowledge retention following the DL phase; the effect of the lag following the DL phase on knowledge retention; and how other training policy and student characteristics are associated with learning and readiness. Performance data from resident training were not available, so we could not test how DL affects readiness for subsequent training for this course. However, we did analyze the association of performance in DL with knowledge retention. As noted above, the sample size for the Battle Staff NCO course was too small to be used in the analysis.

On Average, the Level of Knowledge Retention Was Close to 70 Percent

Figure 3.3 shows the distribution of scores on the Ordnance knowledge retention test. The average score on the test was 69 percent correct (SD = 15, n = 265) (excluding scores from ten students who got a zero on the test), below the typical passing grade of 70 percent. These results indicate that there was sufficient variability in knowledge retention test scores to study learning outcomes. However, the scores of zero, as well as three scores below chance (i.e., less than 25 percent correct) suggest that a number of students were not motivated to perform well on the test. These scores were removed from subsequent analyses.

We created a DL grade point average (GPA) for each student consisting of the average of their scores on four tests administered in the DL course. Each of the four tests was scored on a 100-point scale. In comparison to the knowledge retention test, performance in DL shows much less variability. GPAs ranged from 70 to 97.5, with an average of 85.54 (SD = 5.64).

Figure 3.3
Distribution of Scores on Ordnance Knowledge Retention Test

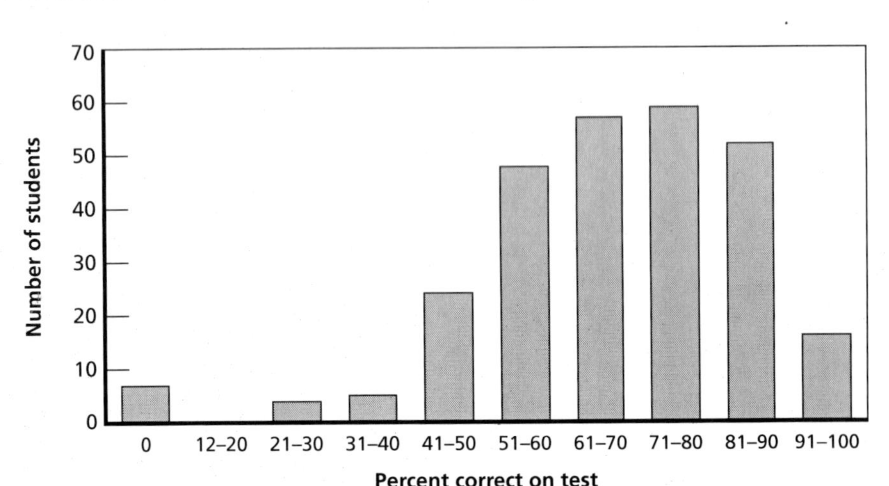

RAND *MG1072-3.3*

Performance in DL Was Associated with Student Cognitive Ability and Job Experience But Not with the Amount of Time Spent on DL

We tested the association of performance in DL with training policy/circumstances and student characteristics using a general linear model.[6,7] Training policy/circumstances consisted of time spent on the Ordnance DL course. Student characteristics included AFQT score, time in the student's new MOS, and civilian experience. Results are depicted in summary form in Figure 3.4. The direction of the association among variables is noted with a (+) or (–) and the strength of the statistical relationship is indicated by the number of asterisks following each variable name, with more asterisks indicating a stronger relationship.

---

[6] The general linear model encompasses a variety of statistical techniques including analysis of variance, analysis of covariance, and regression. These procedures test the relationship between one or more dependent variables (in this case, performance in the DL course or knowledge retention) with one or more independent variables, which here include training policy/circumstances such as time spent on DL and student characteristics such as AFQT score.

[7] Two outliers were omitted from the analyses based on their influence statistics.

Figure 3.4
Learning and Knowledge Retention in Ordnance Course

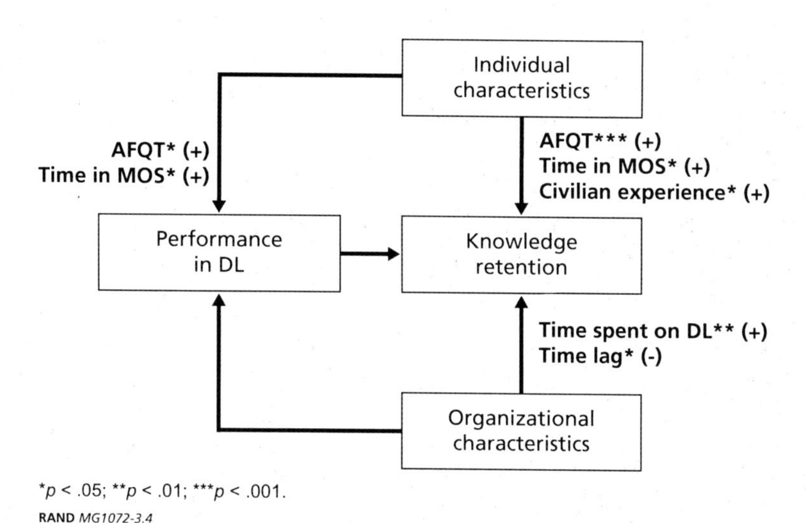

*$p < .05$; **$p < .01$; ***$p < .001$.
RAND *MG1072-3.4*

Results showed, somewhat surprisingly, that the amount of time students worked on the DL phase was not associated with their DL GPA. Students with more experience in their new MOS and students with higher AFQT scores performed better in the DL phase.[8]

Knowledge Retention Was Associated with Time Spent on DL, Lag Time Between DL and Resident Phases, Cognitive Ability, and Job Experience

We also analyzed the association of performance in the DL Ordnance course with knowledge retention. Results show no association between how well students performed in DL and how much knowledge they retained.[9] However, adding training policy/circumstances and student characteristics to the model showed a number of significant moderators of knowledge retention:

[8] For this analysis, $F(9,120) = 2.41$, $R^2 = .15$, $p < .05$. R^2 indicates the percent of the variance in the dependent variable that can be explained by the set of independent variables.

[9] $F(1,129) = 2.71$, $R^2 = .02$, *ns*.

- Students who spent more time working on the DL phase retained more knowledge, controlling for general cognitive ability.[10]
- Students with shorter lags between the time that they completed the DL phase and started the resident phases retained more knowledge than those with longer lag times.
- Students with greater cognitive ability (as reflected in higher AFQT scores) and more relevant job experience (more civilian experience or longer time in their new MOS[11]) performed much better on the knowledge retention test.

Conclusions and Recommendations for Improvement in DL Policies and Procedures

Changes to Training Policies and Procedures Could Improve DL Outcomes

The results of this analysis suggest that there are opportunities to adjust training policies and procedures to improve learning outcomes from DL as well as program efficiency. Some students appear to have completed the DL portion of the Ordnance course too far in advance of resident training or not at all. We observed that 20 percent of the students did not complete the DL phase prior to resident training. In fact, many of the instructors we consulted in planning this study reported that they frequently must use valuable program of instruction (POI) time during resident training to review material from the DL phase.

In addition, some students may not be spending enough time on DL. Course managers reported that many students wait until the last minute to work on DL courses; the students then must "cram" to complete the course (and perhaps these students are especially likely to "click through" the courseware). Numerous studies have shown that cramming, or massed practice, is inferior to distributed practice, par-

---

[10] Controlling for AFQT scores rules out the explanation that students spent more time on the course because they had more difficulty with the material. In fact, there was no correlation between time spent on the course and AFQT scores.

[11] For the full model, $F(11,118) = 8.35$, $p < .001$, $R^2 = .44$.

ticularly for knowledge retention (e.g., see Mumford et al., 1994). In our research, the findings from the Ordnance knowledge retention test suggest that students performed better if they spent more time on DL: students who worked on the course over longer intervals retained more knowledge.

Findings from the Ordnance course also suggest that students with relevant experience may not need to participate in the DL phase of training. We found that students with experience in a major auto dealership service department or industry performed considerably better on the knowledge retention test (with an average score of 86 percent) than did students with light vehicle servicing or "shade tree" mechanic experience (average = 72 percent), who in turn did much better than did students without previous experience (average = 63 percent).[12]

Based on these findings, we offer the following recommendations.

Students should be encouraged to complete the DL phase of the course in a way that minimizes the time lag between the DL and resident phases. Based on results for time lag, policies that permit students to complete DL up to one year prior to resident training are not conducive to knowledge retention. In addition to changing the policy on when students can complete courses, the proponent schools should work to arrange training schedules (e.g., offer duty time for students in particular windows) to enable students to take the DL phase in close chronological proximity to the resident course but without "cramming" DL training into a short time period.

Participation in the DL phase may not be necessary for all students. Exempting qualified students from participation may result in more efficient use of student time and training resources. Data about relevant student characteristics (such as related job experience) could be used as a factor in determining whether some students can place out of all or part of the DL phase or be given a streamlined version of the DL course.[13] Pre-tests could be used to accomplish this goal.

---

[12] $F(1,129) = 28.82, p < .001$.

[13] This recommendation is applicable primarily to courses for which there are opportunities for relevant civilian experience. However, DL prerequisites also might not be needed for students who have already been working in their MOS for a substantial amount of time.

DL courses might be more effective if students were supported in budgeting sufficient time for the course. Although spending more time on online training is not always associated with better learning outcomes (Ely et al., 2008), it is likely that some students in the Ordnance course underestimated the amount of time and effort needed to complete the IMI phase. We recommend sending students a "welcome" message upon enrollment in a DL course that provides guidelines for a timetable to progress through the course as well as contact information for further support. To take a more active approach, instructors could use system logs from the course LMS to identify and contact students who are not making regular progress on DL courses.

We note, however, that more research is needed on the association between time spent on training and learning outcomes. The measure of time in our study, operationalized as calendar days between the first and last course test, is indirect and does not indicate the amount of time that students actually worked on the course. Better estimates could be obtained from system logs that show start and end times for each lesson or module in a course in addition to calendar days (where the latter will indicate whether practice was massed or distributed). Second, it would be useful to measure students' goal orientations, which may influence the association between time spent on training and learning outcomes (Ely et al., 2008).

Other Changes Could Improve the Army's Ability to Evaluate Learning and Training Outcomes

Rather than providing substantive results, the study in the Battle Staff course as well as our experience in the Ordnance course yield a number of lessons learned about evaluating knowledge (or skill) retention and training performance. Many students opted not to take the knowledge retention tests or did not take them seriously. This was more problematic in the Battle Staff course, in which the test was given solely for research purposes. Further, the large number of students who had not completed the DL phase in advance was a serious impediment to evaluating the impact of DL. In both of our knowledge retention tests, we found that barriers to obtaining data reduced the scope of what the evaluations could accomplish.

These experiences lead us to make the following observations to help guide future assessments.

- **Students must be motivated to participate in studies of learning.** Future studies can be more successful if they are conducted as part of an ongoing Army quality improvement (QI) effort (like the process used in the Ordnance course) rather than a research activity in which participation is voluntary. Investigations focusing on large courses can produce lessons learned to guide development of a QI initiative. We envision that regular analysis of knowledge retention would be conducted locally with top-down guidance.
- **Lack of adherence to policy regarding course prerequisites makes it difficult to evaluate the impact of DL.** Enforcing the policy of completing DL prerequisites in advance is important in order to study knowledge retention and performance in the phased approach to training. Of course, completing DL prerequisites presumably is important for pedagogical reasons as well.
- **Increased attention to the quality of course tests is needed.** As noted previously, low scores on the tests could indicate poor knowledge retention, low student motivation, or excessively difficult tests. Test evaluation is needed to ensure that course tests are reliable, valid, and discriminate among high and low performers in the course. There are established statistical methods for evaluating tests, such as Item Response Theory (Lord, 1980), that could be applied within TADLP as part of a QI initiative. These methods also can be used to create adaptive tests, which can provide shorter and more precise estimates of trainee knowledge and skills.
- **A lack of interoperability among systems that contain relevant data poses barriers to evaluating learning.** It can be difficult to access needed systems or data sources that are "owned" by different organizations, and it can be challenging to match students across systems once given access. Those challenges might be somewhat less of an issue if such studies are conducted within the Army where sharing student identifying information (i.e., full

SSN) is straightforward. However, use of these identifiers is more problematic if contractors are needed to participate in evaluation studies—and Army SMEs also expressed concerns about use of SSN as a student identifier, a topic we return to in the next chapter. Even if students can be identified and matched across systems, a lack of interoperability among some systems can also lead to data errors.

We turn to the issue of system interoperability in the next chapter.

CHAPTER FOUR

Feasibility of Using Army Information Systems to Collect Training Evaluation Data

Evaluating training at the program level requires methods to collect and synthesize the data efficiently—in other words, the use of information technology (IT). Thus, a fundamental question in this research is how IT can be used to collect data to evaluate training and to do so in a centralized, standardized way.

We investigated this question by conducting interviews with SMEs in TRADOC Headquarters (HQ), selected proponent schools, and Program Manager Distributed Learning Systems (PM DLS) who are concerned with IT integration issues involving LMSs and the ATIS. We focused primarily on the technical factors surrounding the generation, collection, exchange, analysis, and reporting of training-related quality data, including the development and use of instrumentation and metrics in IT software related to training, as well as the interoperability and integration of training systems and training management systems. We embedded this technical focus in a broader organizational context in order to explore additional issues, such as the kinds of training data that would be valuable, the circumstances in which end-users would support and benefit from exchanging these data, and the factors that would need to be addressed to make data collection, analysis, and integration possible.

This chapter begins with a discussion of our methods. This is followed by a discussion of the key findings derived from the interviews. We conclude with recommendations.

Method

Our primary source of input for this task was a series of semi-structured telephone interviews, augmented by email interactions with contacts in the Army training community. Interviews took approximately 60 minutes. We gained additional input about the current and long-term architectural framework for current and future learning systems from another RAND Arroyo Center project focused on the Army Training and Leader Development (ATLD) Enterprise Architecture. This additional perspective was valuable for identifying and analyzing interoperability issues that underlie the effective sharing and analysis of training quality data.

Drawing on our team's past interactions with staff from TRADOC, ATSC, and other parts of the ATLD community, we developed a list of candidate interviewees whose experience and opinions seemed most likely to be relevant to our study and who represented a broad range of groups, including ATSC and TRADOC decision makers, instructional system designers, training developers, course managers, and others in ATSC, TRADOC HQ, selected proponent schools, and program offices for relevant training IT projects. This candidate list evolved over the course of the study, as interviewees suggested other possible interviewees. The eventual list of candidates included 93 names, all of whom were invited to participate. Of these, about 50 responded, and 35 of those expressed willingness to be interviewed. Due to scheduling constraints, 25 interviews were conducted.

SMEs' roles and backgrounds were often multifaceted: many were responsible for performing multiple roles in their current position or had held a previous position that involved varied experiences, all of which informed their responses. SMEs' primary current responsibilities were as follows: 5 represented TRADOC HQ, ATSC, or Army-level organizations; 12 represented TRADOC schools (course managers and course developers); and 7 represented organizations that develop and maintain IT systems or infrastructure (such as program offices or contractors).

Interview topics included the following:

- Interviewee's role in designing, developing, managing, or using training systems.
- General questions about the value of and methods for automatic collection of training data, including system logs, student surveys, and SME evaluations; value of instrumenting course development and delivery tools.
- Questions about the systems that interviewees use, including the kinds of data that are collected and analyzed; technical features of the systems (e.g., evaluation and reporting capabilities, interoperability with other systems, data standardization); methodological considerations, such as types of analysis supported; organizational/cultural issues surrounding collection, analysis, and sharing of training evaluation data; and questions about the LMS in use.

The specific questions for each topic are presented in Appendix I. As is often the case in interviews, not all respondents replied to all questions, and some emphasized issues that were not included in our questionnaire. Such excursions from the script can sometimes be among the most useful inputs in a study, since they may reveal issues whose importance might otherwise not have been recognized. Our analysis is based on all relevant input received, whether or not in direct response to any of our questions.

To analyze the data, we collected and combined related answers in order to identify the key points relevant to each question. We then analyzed and integrated these points across the various questions and categories to look for common threads and significant issues. Finally, we considered these issues in terms of our previous and ongoing experience related to ATSC and TRADOC projects.

Findings

In this section we synthesize and summarize the main issues raised by our interview participants. Throughout this section, we intersperse our own comments and reactions with some of these responses.

We first discuss SMEs' views of the value of collecting training evaluation data. Next, we present technical and nontechnical barriers to collecting, analyzing, and integrating evaluation data.

SMEs See Value in Evaluation Data

SMEs discussed two general categories of evaluation data: student-level data and course characteristics. Student-level data include cognitive leaning and training performance (performance on knowledge and skills tests, respectively), usage statistics (e.g., time spent on courses obtained from the LMS), and student reactions (e.g., responses to end-of-course surveys). Course characteristics include attributes of courses, such as IMI levels, delivery modes, type of developer (in-house or contractor), and graduation rates.

SMEs generally felt that gathering, sharing, and analyzing DL training evaluation data, if done well, could have significant value to schools, course developers, ATSC, and possibly commanders and students. Most respondents felt that sharing data among schools might create a useful exchange of ideas and best practices for producing effective DL. A majority also felt that sharing data upward from schools to ATSC could help rationalize and justify the need for particular courses, align POIs with current doctrine, and lead to a better understanding of the effectiveness of various online course development and delivery techniques.

SMEs felt that course managers could use student-level data for several purposes. Assuming that the data indicate that students take these courses and derive some benefit from them, student-level data could be used to document the value of DL courses and to justify the need for training resources, such as mobile training teams (MTTs), more course developers or instructors, or funds to develop or improve training systems. Such justifications are currently difficult due to the lack of quantitative evidence as to the value of any given course. Because online courses are relatively new and are different from traditional courses, respondents felt that evaluation data about DL would be especially valuable.

SMEs also reported multiple benefits from capturing course attributes:

- Capturing IMI levels would be valuable in documenting the time needed to develop DL courses utilizing IMI or to evaluate the effectiveness of using those levels.
- Measuring the percentage of learning that is mobile would either verify or refute the current perception that most training is still done in classrooms. The percentage of mobile training should be measured in terms of both the relative number of classroom versus nonclassroom hours of instruction and the relative number of phases and modules that use each mode (e.g., traditional classroom, MTT, VTT, IMI, etc.). Linking the delivery mode to measures of training quality would enable courseware to be designed on the basis of training performance rather than on the basis of technology, as is now often done.
- Recording and aggregating data about which courses are developed in house (versus being outsourced) could be used for resourcing decisions and to justify increased flexibility to allow proponent school staff to choose among in-house authoring, local contracts, master contracts, etc. Such information would help balance this flexibility against oversight by giving ATSC visibility into the proportion of funding that is being administered by each type of contract.
- Downward sharing of aggregated course and student performance data (such as performance on tests, student reactions, and graduation rates) from TRADOC could also enable schools to improve their allocation of resources to DL versus classroom courses. It would also better enable them to schedule and coordinate their use of training bandwidth in light of Army force generation (ARFORGEN) events (such as deployments) that might have competing and overriding demands.
- Capturing and analyzing data about the types of contracts and oversight mechanisms used to develop specific courses, combined with their development times, might help identify ways to shorten the current average contractor-based two-year course development process. Such a reduction would be of particular value: given that the doctrine revision cycle is about 18 months, the development

cycle must be shortened to prevent the creation of outdated course content.
- Sharing administrative data, such as course enrollment and graduation rates (see also Shanley et al., forthcoming) may have value to some stakeholders. SMEs felt that commanders want to know how many students have been trained in particular skills, but are less interested in how individual students perform than in how courses perform.

Training Evaluation Requires Data from Multiple Sources

Many of the types of analyses just described would require data from multiple sources. For example, the assessment of knowledge retention that we presented in Chapter Three required the following data:

- Test scores from DL training, which were provided by ATSC (but which might in the future be obtained from the course LMS).
- Test scores from resident training, which were provided by the course manager.
- Demographic characteristics, which were obtained from TAPDB and RCCPDS.
- Registration, enrollment, and graduation data, obtained from ATRRS.
- Knowledge retention test scores (from a RAND server—but which, again, may come from an LMS in future studies).

Some variables used in the analyses were not available directly but had to be derived by a programmer (e.g., time spent on tests or time between the completion of DL and the start of resident training). Likewise, some of the potential studies of course characteristics identified in our general model in Chapter One, or by SMEs (e.g., linking delivery mode or IMI levels to training outcomes), require access to data maintained in different locations and from databases kept by schools or ATSC.

Our interviews revealed a variety of impediments to obtaining data needed to evaluate training. We grouped barriers into technical and nontechnical categories. Technical issues involve ATLD system limitations and the mechanisms that enable them to interoperate with each other and to interact with their users. Nontechnical issues involve methodological, organizational, and policy factors.

There Are Numerous Technical Impediments to Obtaining Data from Different Information Systems

SME interviews indicated that the current state of the information systems that maintain training data poses significant barriers to evaluating training at the program level. We first discuss how a number of system and usability problems affect collection of evaluation data. We then discuss interoperability, which appears to be the most significant technical impediment to sharing and integrating evaluation data.

System and usability problems threaten validity of data. Course delivery systems and LMSs can be subject to serious technical problems, including system crashes or disconnects that require students to log in repeatedly or that lose the results of a session that has not yet been completed. For example, several respondents noted that, due to access problems or bugs, Saba (the LMS system software) sometimes reports false "fail" events to ATRRS for students who have actually passed a course. A related problem involving user interface design is that a user's session can sometimes end unexpectedly when the user is attempting to close the Saba window. Issues related to underlying interoperability problems with LMSs can prevent a student's completion of a course from being recorded correctly in ATRRS.

Technical problems frustrate students and interfere with the learning process. They also affect training evaluation by introducing artifacts, e.g., by making it appear that a student has taken longer to respond than he or she really has or by recording incorrectly that a student has not finished a course. As a result, data pulled from systems such as ATRRS and LMSs do not accurately reflect DL course usage. Technical problems might also influence student reactions by negatively biasing ratings of other aspects of the course, such as the quality of course content. Proponent school staff identified the lack of sufficient network bandwidth and reliable access to remote servers as one of their main problems. The lack of infrastructure limits the usability of online courseware for their students and undermines the training process. Similarly, technical problems involving bandwidth or speed were the main issues reported by nongraduates (see Chapter Two).

Interoperability among training data systems is limited, ad hoc, and error-prone. Interoperability is the most challenging technical problem for evaluation. Currently, the data needed for training evaluation, such as student or population data (e.g., demographics, assignment history, test scores) and courses data (e.g., POI), are typically not available from a single source (and in some cases, may not be available at all). Therefore, to conduct evaluations it is necessary to find and query multiple data sources, most of which are not connected to each other and so require specific expertise or authorization to access and use. Furthermore, identifying and joining data across multiple sources can be problematic because each system may define and encode data in unique or idiosyncratic ways. For example, systems may not use the same student or course identifiers or use identical semantics on their data (such as the definition of a "training module").

Respondents cited numerous cases in which finding, accessing, and interpreting all of the data they needed to perform meaningful analysis were difficult or error-prone using existing ATLD databases and systems. As described in Chapter Three, we encountered similar problems in performing our study due to inconsistencies in student identifiers based on names across different databases. Our problems were only partially resolved by the intensive efforts of a research assistant—efforts that would not be practical as a routine solution.

Some of these systems are connected to others using unique, pairwise (i.e., one-to-one) interfaces, but these connections are often difficult to maintain. For example, ATRRS has over 40 such pairwise interfaces, and many of these were built using now-obsolete techniques that must still be maintained because the system on the other side of the interface has not evolved beyond its original technology. Many other systems are not linked at all and require "hand-jamming" of data (e.g., printing data from one system and manually reentering it into another). Some systems that are referred to as databases are really just text repositories. For example, much of the POI and Course Administrative Data (CAD) information in ASAT[1] is in text format, which cannot easily be used to conduct queries needed for data analysis.

Although new pairwise interfaces can be forged between systems when necessary, this process typically requires formal agreement between program offices, schools, contractors, or vendors, along with the allocation of funds and programming resources to engineer and test each new interface. In addition, pairwise connections of this kind do not scale well as more systems are connected, since each such interface tends to be unique and must be maintained over time as systems evolve independently.

Several SMEs emphasized that interoperability requires more than just technical interoperability. It also requires compatibility among the names used and the meanings of corresponding data elements (known as *semantic compatibility*) and policy alignment (e.g., conformance to regulatory policies on security, privacy, and information assurance). Respondents noted that although the proposed service-oriented architecture (SOA) approach (which we discuss later in this chapter) provides mechanisms for describing semantics and policies, it does not automate the process of creating such descriptions or interpreting them correctly when invoking SOA services. Similarly, SOA does not by itself solve problems such as the lack of a single sign-on to Army training systems, inconsistency in the use of email addresses and other student identifiers, and the dynamic nature of course names and designations across the Army.[2]

There Are Also Nontechnical Barriers to Collecting, Analyzing, and Aggregating Training Evaluation Data

Interviews with SMEs revealed a number of methodological, organizational, and policy impediments to evaluating training. These concerns were focused more on the concept of program level evaluation than on use of IT systems per se.

---

[1] ASAT, or Automated Systems Approach to Training, is the information system for management of Army training products.

[2] Single sign-on allows users to log in once and access multiple systems rather than having to log in to each system separately (often with a different login and password).

Aggregate measures of training outcomes are not comparable across schools and may be subject to misinterpretation. For example, course completion rate is a gross measure that provides a poor basis for evaluating attrition due to any specific substantive cause—and may be due to record-keeping artifacts, as documented in the nongraduate survey reported in Chapter Two. In theory, training evaluation measures such as pass/fail rates or number of attempts on tests should be comparable across courses, but aggregating these outcomes can also be subject to misinterpretation. For example, high failure rates on knowledge tests could be indicative of bad tests, poor course content, course delivery problems, or poor student performance. Although test evaluation could rule out some of these explanations, few schools evaluate the quality of course tests. The meaning of low-level usage statistics, such as time spent on a lesson or module, also can be ambiguous given the myriad technical artifacts in online courses discussed earlier. This makes human analysis crucial in order to reduce misinterpretation of results.

We note, however, that the development of more meaningful metrics of student interaction with online training tools is an area of active research, not just in the Department of Defense (DoD) but across a broad range of training and education domains. The current study, as well as other research efforts, should produce a better understanding of interaction effects, artifacts, and performance in online training. In addition, the development of commercial Web-based interfaces is producing improved tools for measuring user interaction. For example, scripting tools such as Asynchronous JavaScript and XML (AJAX) allow browsers to capture information about user keystrokes and the time spent looking at particular content. Similarly, artifacts due to factors such as system crashes should decrease over time as system and network reliability improve and servers become more appropriately distributed and available. Whereas there may always be limits to how reliably low-level interactions can be interpreted, greater awareness of these limitations should yield a better understanding of what can and cannot be inferred from such data.

Numerous respondents felt that in order to measure the impact of training content, design, or delivery, it would be necessary to evaluate not just student usage and training performance but also the association of these outcomes with relevant job performance or training transfer. At the same time, many respondents were skeptical about the feasibility of defining meaningful measures of this kind as well as actually measuring them. We believe that defining meaningful measures of job performance *is* feasible, as there are well-established measures of performance in many types of jobs, although we agree that collecting these data may be difficult (see Chapter Three). Furthermore, establishing the link between training performance and job performance is even more challenging, given difficulties in collecting these data (see Chapter Three) and the many factors other than training that can influence performance on the job. Measuring training outcomes such as cognitive learning and skills and post-training attitudes such as self-efficacy can be useful indicators of training quality even in the absence of measures of job performance.

SMEs were concerned that they may be asked to collect data without a clear rationale. SMEs' responses regarding the validity of aggregating training evaluation data are relevant to another one of their concerns, i.e., that proponent school staff will be asked to collect data without adequate rationale. Several respondents noted that measurement, data collection, and analysis could be a "black hole" that consumes resources without any legitimate reason, and some commented that there is a long tradition of over-reporting in the ATLD, Army, and DoD communities. However, the consensus among respondents was that schools would be willing to share such data if they received some direct benefit from doing so, such as the ability to improve the quality of their courses or reduce development time.

SMEs were also concerned about how shared data would be used. SMEs also have questions about how training evaluation data would be used. Most training evaluation data collection is performed by schools, and although some schools share data with each other, the organizational culture is one in which schools generally consider course performance measures to be their own concern. In some cases, respondents also cited turf issues or bureaucracy as impediments to data sharing. More often, however, respondents saw a significant risk in sharing their data upward (with higher HQ), since they worry about losing resources. For example, school staff fear that data will be used to shorten a course, which in turn will reduce their resources for administering and delivering that course.

Data collection may strain personnel resources and skills. Some types of data needed for evaluation are currently not captured—or are not done so automatically—in current IT systems. Therefore, human effort will be required to collect, enter, and/or process data. For example, data concerning a student's task performance in an assignment following training typically would need to be generated by the student's superior. Other outcomes can be collected automatically but require effort for data synthesis (e.g., computer programming to generate a course GPA). In some cases, existing tools can collect relevant evaluation data, but because the tools are administered by contractors, course managers or other ATSC or TRADOC personnel must ask the contractor to generate a report to obtain the desired information. This is seen as too cumbersome (and sometimes costly) to serve training analysts and decision makers, who often need to perform ad hoc queries to address specific questions or problems.

Similar issues were raised about collecting data about course characteristics, such as IMI levels. Although some schools already collect information of this sort in POIs, which are available from ASAT, some respondents noted that such data may be misleading because IMI requirements in DL contracts may not reflect the IMI levels produced. In fact, in our previous work, we found that most modules used IMI Levels 1 or 2 even though the contracts specified Level 3 (Straus et al., 2009). In addition, information about IMI levels is generally provided in text format and so is not easy to extract automatically.

Thus, any data that rely on human input or manipulation will be complete, accurate, and consistent only to the extent that the human (or organization) in question has sufficient resources to collect the information and is appropriately diligent. Respondents pointed to a number of cases (as described throughout this chapter) of incomplete or inaccurate data in existing systems, due at least in part to a lack of suitable incentives and resources for the personnel who must process the data. (Other causes are a lack of data-checking within systems and a lack of interoperability among them.) A related concern is whether available staff in the Army training community have the expertise required to analyze evaluation data. Responses to this question were split: roughly half the respondents felt that this is not a problem, whereas the other half felt that it is, at least for ATSC.

Existing policies on data sharing are insufficient. Finally, SMEs generally agreed that policies are needed to promote sharing and analysis of evaluation data. Despite the tendency for schools to keep training evaluation closely held, most agreed that ATSC, TRADOC, and DA should be able to get any data they need. Some felt that existing policies are sufficient for this purpose if they are interpreted correctly and are enforced, whereas others felt that new policies are needed. Some also noted that commanders often tailor policies to their own priorities, which do not always reflect those of the larger ATLD or Army enterprises. In addition, many respondents believe that policies, whether existing or new, are not enough by themselves and that top-down guidance and incentives are needed to support a cultural shift in order to overcome the tendency for schools to keep data to themselves.

SMEs also discussed the need for policies that address privacy issues. In general, respondents felt that aggregated data should be free of Personally Identifiable Information (PII), though some felt that demographics could be a problem, even if data are aggregated (e.g., in cases where a specific subset of the population is so small that individual identities can be inferred). Disaggregated course grades are considered protected information. A number of respondents felt that data should be de-identified before aggregation, citing early incidents in which the Army Learning Management System (ALMS) exposed some students' social security numbers. We note, however, that any matching of student records across databases must be conducted before data are de-identified.

Conclusions and Recommendations to Address Technical and Nontechnical Barriers

Although the SMEs in our interviews see value in collecting evaluation data, they also identified a number of technical and nontechnical impediments. Key technical barriers include system and usability problems and, especially, poor interoperability among training information systems. Methodological, organizational, and policy issues further impede the collection, analysis, and sharing of training evaluation data. Among the issues cited by SMEs were a lack of comparable measures across schools, unclear rationale for collecting data, and insufficient personnel resources to support data collection and analysis. These concerns are consistent with reasons for resistance to training evaluation in organizations more broadly (Salas and Cannon-Bowers, 2001).

We offer the following recommendations to address technical and nontechnical impediments.

Addressing Technical Impediments

Achieving interoperability is a long-range solution that will require substantial effort and resources. Therefore, we present short-term, medium-term, and long-term recommendations to enable training staff to obtain data needed to assess training effectiveness while moving toward system interoperability.

We have three recommendations to help schools enhance their ability to evaluate training in the short term.

Move to online administration of tests and surveys. Schools can enhance their ability to evaluate training by moving to online administration of tests and surveys—a function supported by most courseware development and delivery systems. Clearly, digital capture of these data will facilitate a wide range of analyses and will ease reporting. Digital capture will also preserve data that are frequently discarded, such as responses to individual items on tests, which are needed for test evaluation. Online administration can also enhance test security, for example by scrambling items or item response options or by administering alternative versions of a test. Some online delivery systems also have instrumentation to capture low-level interaction data such as page-visit frequencies, dwell time, etc., or they can be programmed to collect this type of data. We recommend that training staff make use of these system features.

Collect different types of data within a single instrument. A second recommendation is to collect as much data as possible in a single instrument in order to obviate the need to pull data from different systems. As one example, surveys or course tests can include questions about student demographics. Although this approach is a workaround to accessing data directly from their original sources, it may eliminate or reduce the need to obtain data from systems such as TAPDB or RCCPDS.

Continue manual studies. There is no need to postpone all evaluation until IT systems can fully support them. Manual studies may take longer, but in addition to providing results about the quality of training, they can yield lessons learned for designing automated collection of evaluation data.

We have three recommendations for the medium term to help build the Army's ability to collect training evaluation data.

Develop standards for evaluation data. TRADOC HQ recently redefined the ALMS to include other LMSs, i.e., Blackboard and AtlasPro, in light of some schools' need for functionalities that these alternative systems provide. However, expanding ALMS in this way should not supplant attempts to improve consistency and interoperability among LMSs. Thus, the training community should begin the process of developing standards for evaluation data in order to produce (among other benefits) more complete, consistent, and available data. Such standards are also necessary to move to SOA.

Modify training systems to consistently collect data. Once standards have been determined, systems should be modified accordingly. This process does not have to be completed all at once, but can be conducted incrementally, starting with systems and data elements that will have the greatest impact.

Develop Web services to allow database queries. ATSC also should investigate Web-services interfaces to allow database queries in the interim. For example, Web-services interfaces to ATRRS or ASAT would enable systems that do not already have pairwise connections to these systems to query their data with a minimum of programming effort. Contractor-administered systems should be included in this effort.

We have two recommendations for the long term.

Move to service-oriented architecture (SOA). Recently, organizations in all sectors have begun to seek greater efficiency, effectiveness, and coherence through increased integration of their activities at the enterprise level. This has placed a new burden on existing IT systems to interoperate with each other to support integrated, enterprise-wide functions. Meanwhile, new systems are increasingly being built with current and future interoperability as an important design criterion. In accordance with this widespread need, new architectures, notably SOA, have begun to emerge to support interoperability. Although SOA is not a panacea and introduces its own challenges and limitations (as discussed in Appendix J), it is highly attractive as an interoperability enabler.

Adopting SOA as the "to-be" architecture for ATLD appears to have the potential to greatly facilitate the collection and sharing of evaluation data, at least in the long term. Since both recent Army Training Information Architecture (ATIA) and emerging ATIS efforts (as well as Army and the DoD as a whole, via the Global Information Grid) have adopted SOA, this may constitute a viable evolutionary path toward the improved collection and sharing of online training evaluation data.

SOA repackages the capabilities of systems into a set of "services," each of which runs on some server on the network and obeys a set of protocols that enable users and other services to find it and invoke it dynamically over the network. Any other service, system, or user can then use any combination of these services over the network, discovering and invoking them by means of the SOA protocols, to perform a desired business process. (See Appendix J for a more detailed discussion of how SOA operates and its advantages, limitations, and challenges.) In principle, SOA would enable training systems to interoperate with each other without prior agreement of any kind, thereby bypassing the cost and lead-time of creating specialized pairwise interfaces between them. Any training system whose functions are available as an SOA service can be discovered and invoked dynamically by any other such system. This would facilitate the exchange, sharing, and analysis of evaluation data among multiple training systems, support the creation of "training dashboards" for a variety of users, and provide ad hoc query capabilities that decision makers could use to answer new questions related to training as they arise.

We recommend that ATSC and TRADOC proceed incrementally toward implementing an SOA architecture for ATLD systems. Incremental progress toward SOA can be made by first converting "key" systems (possibly including ATRRS) and those that interface with them, while converting less highly connected systems later. This approach would require maintaining existing versions of systems available for use by other legacy systems that will not yet have been converted to SOA at any given time. An organization within the ATLD community would have to take responsibility for directing and coordinating this process, and funding would have to be provided for converting existing systems or creating new ones. ATIS may form the basis for this incremental SOA development and should in any case provide tools and technical support to facilitate it.

As part of this move toward SOA, the ATLD community should develop an ATLD ontology to explicitly and formally describe, represent, and harmonize the semantics of training-related terms, database elements (names and meanings), input and output parameters of training-related systems, and the processing performed by these systems. This is a necessary precondition to implementing SOA and indeed an important step in fostering automated interoperability among ATLD systems, whether or not they use SOA.

Develop a training dashboard. Once the Army achieves improved data standards and system interoperability, ATSC should consider developing a training dashboard as a means for collecting, analyzing, and/or disseminating training evaluation data. The idea of a training dashboard generated quite a bit of interest among SMEs in our study, although they had varying views of its meaning and uses. A training dashboard can be used as a low-level tool to enable students (or their instructors or supervisors) to assess where individual students stand in a course; as a mid-level tool to enable schools to track course usage, course characteristics, student performance, student population data, and information about facilities, devices, and ranges needed in some courses (see Figure 4.1); or to provide a Common Training Picture

Figure 4.1
Mock-Up of a Training Dashboard Displaying Course-Level Data

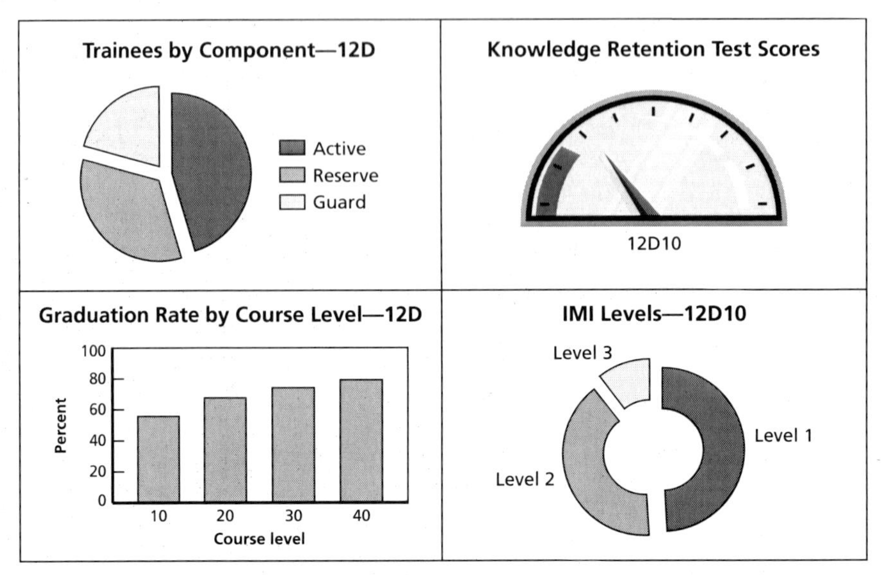

RAND *MG1072-4.1*

(CTP) for high-level decision making. Determining what information to include on a training dashboard and identifying methods to "roll up" the data from low to high levels requires careful analysis to ensure that aggregated data are valid, meaningful, and actionable.

Addressing Nontechnical Impediments

In this section we present recommendations to address methodological, organizational, and policy issues identified by SMEs. The recommendations are based on the input of SMEs and our own analysis. The goal of these recommendations is to design an enterprise-wide program of evaluation in which data-collection efforts provide clear value to stakeholders.

Build end-user participation into all phases of process design. We recommend inviting staff from proponent schools and centers as well as other organizations affected by potential changes to participate in all phases of designing processes to collect and evaluate training data. Involving stakeholders is an essential first step, both because end-users have the technical and organizational knowledge and skills to know what will and won't work, and because participation will enhance buy-in to process changes. We recommend the use of ongoing workgroups for each of the steps outlined below and to vet proposed changes to the Army DL community.

In addition to participating in process design, proponent schools should be able to customize evaluations to address local needs. They should also have the opportunity to review program-wide reports in order to mitigate potential misinterpretation of results.

Establish the business case before requiring any new data collection. A second recommendation is to evaluate and communicate the business case for collecting, analyzing, and aggregating data at the enterprise level. These efforts would be undertaken to determine the value and feasibility of collecting various kinds of training evaluation data and to communicate the rationale for enterprise-wide evaluation efforts. The business case should establish the return on investment for these efforts and make clear the direct benefits to schools in terms of contributing to improvements in course quality or efficiency in course development.

Develop appropriate policies to support evaluation. Policy development is needed with respect to several topics. First, the business case should be used to establish data-reporting requirements. These requirements should be crosswalked and integrated with relevant existing policies to avoid requiring over-reporting.

Second, policy should be developed that spells out how training evaluation data will be used at the enterprise level. For example, under what circumstances would results be used to decrease the length of a POI or to merge or eliminate courses? We recommend using a shared governance process to make these decisions and to prevent unilateral decisions to reduce school resources on the basis of evaluation results.

Third, policy is needed to determine how data will be reported in order to address privacy issues. Policies regarding privacy should be developed in coordination with the G-2, G-6, and other relevant offices. Among other issues is the use of SSNs as a training identifier. An alternative is the electronic data interchange personal identifier (EDIPI), which is embedded in all Common Access Cards (CAC). However, this number cannot currently be used by all soldiers to register for courses because not all computers have CAC readers.

Provide requisite resources and incentives. ATSC should ensure that the proponent schools and centers have the resources they need to collect, analyze, and report evaluation data. This should include providing hardware and software for collecting, analyzing, and/or reporting data, resourcing the personnel needed to support these efforts, and providing training in analytical techniques. ATSC should also encourage, reward, and disseminate best evaluation practices.

Establish an evaluation data analysis organizational entity. Finally, we recommend establishing an entity to support enterprise-level training evaluation. This office would:

- Provide analytical support to proponent schools and centers.
- Identify relevant interoperability shortfalls and serve as a liaison to coordinate data exchange.
- Coordinate data-collection efforts across the schools.
- Integrate results (with input from schools) and report to ATSC and DA.
- Collect and disseminate lessons learned and best practices.

To summarize, it appears feasible and desirable to use Army IT systems to collect both student-level DL training evaluation data and DL course attributes. If done properly, systematic collection of such data could have significant value to a variety of stakeholders including ATSC, schools, training staff, commanders, and students. In addition to improving course design, delivery mechanisms, and training outcomes, such evaluation could help measure and justify the true cost and value of DL. Technical obstacles to obtaining useful and meaningful data include learning system access and usability problems as well as the current limited, ad hoc, and error-prone interoperability among training data systems. Nontechnical obstacles include the lack of standard, comparable measures of training evaluation across schools and concerns that shared data might be misinterpreted or used as an excuse to reduce training resources. Moreover, a convincing business case must be made for adding new data-collection features to existing systems in order to justify any added burden this would place on system developers and users. The short-, medium-, and long-term recommendations offered above address the identified technical obstacles and provide a number of procedural and policy recommendations to address the associated nontechnical obstacles. A coordinated effort along these lines should result in improved training evaluation data that are shared among schools and TADLP, leading to the more effective use of DL, more accurate training and course data, better situational awareness of training performance, and improved training outcomes.

CHAPTER FIVE

Conclusion

Evaluation is needed within the Army's DL program to improve the quality of DL courseware and training processes, to increase DL utilization and efficiency, and to help the program compete for resources to meet program goals. The studies described in this report built on previous RAND Arroyo Center research dedicated to the design and testing of tools for conducting training evaluation (e.g., Shanley et al., forthcoming; Straus et al., 2009). For this research, we developed and tested new tools and methods for evaluating Army DL, identified factors associated with training quality, and provided direction for improvement in TADLP. More specifically, we:

- Presented a model for evaluating DL within the Army that can guide future assessments and serve as a basis for the evolution of a training evaluation strategy within TADLP.
- Designed and piloted two survey tools that provide program-level feedback from students and that can be used for ongoing evaluation in a wide range of DL courses.
- Designed and partially tested a method to measure DL's contribution to student learning, knowledge retention, and readiness for subsequent training.
- Determined how IT systems might better support data collection for evaluation of DL.

In the remainder of this chapter, we highlight key findings and provide recommendations for future research and practice.

Key Findings

The results of this series of studies and assessments provide information on the current state of DL and suggest that a more comprehensive program of evaluation, better supported by the Army's IT systems, could provide major benefits to TADLP.

The research led to some important substantive findings about the DL program. In brief:

- The nongraduate student survey indicated that DL graduation rates were substantially higher than those derived from ATRRS, and most of the reasons for nongraduation were found to be due to factors that are outside the DL program. At the same time, results point to ways to increase graduation rates.
- Moreover, the graduate survey showed that students who did complete DL courses were moderately satisfied with multiple aspects of the quality of DL instruction and support. The areas with the greatest need for improvement were technical issues related to bandwidth and the degree to which the courseware was engaging to students.
- The assessment of knowledge retention revealed the importance of spending sufficient time on DL and of minimizing the lag between DL and resident instruction in the phased approach to training.
- Interviews with SMEs pointed to technical and nontechnical barriers to data collection and analysis, including system and usability problems, poor interoperability among training information systems, a lack of common measures for assessing DL, and insufficient personnel resources to support data collection and analysis.

All studies within this project led to recommendations for changes in policies and procedures that could improve Army DL outcomes. For example, our nongraduate survey indicated the need for improved administrative support for students following enrollment and better record keeping regarding student status in DL courses. Our graduate survey suggested a need for greater learner engagement with IMI content and revealed that students want substantially more interaction with instructors and peers in DL courses. Our studies of learning indicated that evaluating knowledge retention can provide substantive results relevant to course development and to training policy and procedures. Our assessment of IT capabilities led to recommendations to enhance the Army's capability to collect training evaluation data.

These studies also demonstrate the feasibility of and methods for conducting DL evaluation within the Army:

- Implementation of the surveys showed that structured student feedback can provide useful input for improvement in the DL program. From a methodological standpoint, the survey questions are relevant to a wide range of courses, measures are psychometrically sound, and the surveys are not burdensome to answer.
- The studies assessing knowledge retention revealed that evaluations of cognitive learning and soldier readiness for subsequent training are feasible within the Army. The studies also provided direction on methods for conducting more successful evaluations in the future.
- The model, tools, metrics, and methods used in this project can be used to evaluate other forms of Army training with minimal revision.

A summary of recommendations for training evaluation and for DL design, implementation, and policy is provided in Table 5.1; comprehensive descriptions of recommended changes are provided in Chapters Two, Three, and Four.

Table 5.1
Summary of Key Recommendations

Study	Key Recommendations
Student Surveys	**Training Evaluation** • Improve capture of information about student status in courses by documenting student purpose, updating course graduation field in ATRRS, and notifying students who do not start a course within a specified time period. Consider requiring documentation of graduation for prerequisite courses in ATRRS prior to registering for subsequent phases of a course. • Make the graduate and nongraduate surveys part of routine QI efforts. • Use a common set of survey indicators across schools, while allowing schools to add items addressing course-specific topics. • Provide a common platform and software application to enable schools to design, administer, and score surveys.
	DL Design, Delivery, and Policy • Improve technical features of DL, including bandwidth and access. In short run, provide low-bandwidth versions of courses or CD-ROMs for students in bandwidth-constrained settings, or break longer lessons into smaller sections and eliminate nonessential data-intensive graphics. • Improve courseware design to increase student engagement. • Design and implement DL in ways that provide greater opportunities to interact with instructors and peers. • Enforce policy of allowing soldiers duty time for required training; consider providing educational duty hours (EDY).
Knowledge Retention	**Training Evaluation** • Incorporate future studies into Army QI efforts. • Enforce policy of requiring students to complete DL prerequisites in advance. • Assess reliability and validity of course tests.
	DL Design, Delivery, and Policy • Encourage students to minimize the time lag between completion of DL and resident phases of a course. • As appropriate, allow students with relevant course experience to skip the DL part of courses. • Monitor system logs for progress through DL courses, and assist students in budgeting sufficient time for completion.

Table 5.1 (continued)

Study	Key Recommendations
SME Interviews	**Technical Impediments**
	• Move to online administration of tests and surveys.
	• Continue manual studies; collect different types of data within a single instrument.
	• Develop standards for evaluation data.
	• Modify existing systems to consistently collect data.
	• Develop Web services to allow database queries.
	• Over time, move to service-oriented architecture.
	• Develop a training dashboard.
	Nontechnical Impediments
	• Build end-user participation into process design.
	• Establish the business case for new data collection.
	• Develop policies to support data collection, use, and reporting.
	• Provide requisite resources and incentives for data collection, analysis, and reporting.
	• Establish an enterprise-level training evaluation entity.

The Suggested Way Ahead for Army Evaluation of DL

Move to Widespread Implementation of Student Surveys

Some of the tools we developed are ready to be turned over to the Army for implementation. Widespread adoption of student surveys would help individual schools and centers to systematically monitor aspects of the quality of their courses and would enable ATSC to obtain measures that could be aggregated in any number of ways to report on program quality. In addition, relatively little modification would be needed to use the surveys for other forms of DL, such as blended learning or MTTs.

The surveys should be updated on a regular basis to maintain their currency and to address new issues as they arise. For example, the surveys can address other impacts of DL, such as whether it enhances or detracts from soldiers' opportunities to spend time with their families. As recommended in Chapter Two, other questions could be added to fit the needs of particular schools.

Expand on Current Research Efforts Focusing on Learning

Understanding the extent to which students learn and retain knowledge is critical to documenting DL's contribution to readiness. Whereas the DL experience surveys are ready for immediate use, evaluation of learning from DL (e.g., using knowledge retention tests) needs additional investigation. Future studies of student learning should be conducted in large courses with complete data from both DL and resident training phases in order to test the impact of learning on individual soldier readiness. Moreover, an evaluation would have even greater value if it focused on a course in which DL material also supports operational training by serving as a job aid; such a focus could yield valuable lessons and raise awareness of DL's value to the Army training community more generally. We anticipate that these investigations also would set the stage for measuring the impact of training on job performance.

As discussed in Chapter Three, successful evaluation of student learning requires careful attention to the quality of course tests. Thus, another possible expansion of current research is to evaluate a broad sample of DL tests. Such research could be used to demonstrate how to assess whether test questions are at the appropriate level of difficulty and whether they adequately discriminate between good and poor performers.

Conduct Evaluation Studies Using a Broader Set of Measures

Measuring student reactions and cognitive learning is only part of a comprehensive program of evaluation. As shown in the integrated model of training evaluation described in Chapter One (repeated here for convenience as Figure 5.1), there are many measures of training evaluation, and multiple measures can complement each other. For example, analyses of graduation rates did not tell the whole story about usage of DL courses without surveys of nongraduates. Measures to address in future studies are shown in italics in the figure.

Figure 5.1
Integrated Model of Training Evaluation and Training Effectiveness

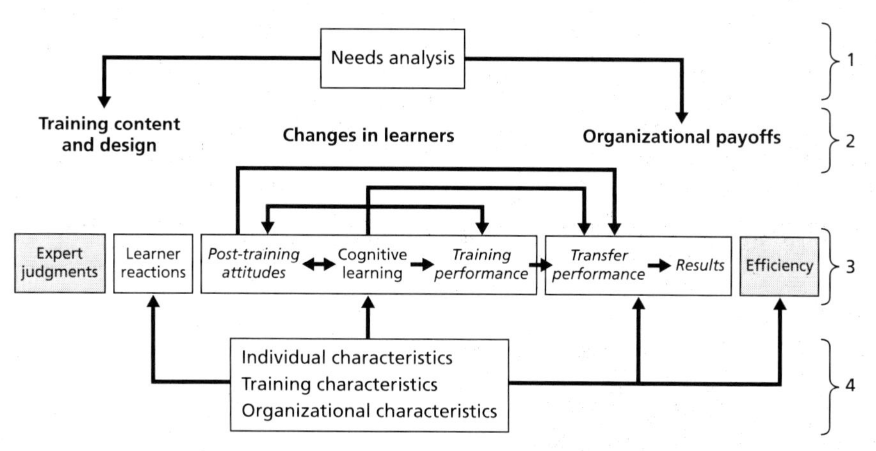

SOURCE: Adapted from Alvarez, Salas, and Garofano (2004), p. 393, Figure 1. Used with permission.
RAND *MG1072-5.1*

One important training evaluation measure is training performance, i.e., acquisition of skills. As described in Chapter Three, the present research deliberately measured declarative knowledge (cognitive learning). A large proportion of Army training, however, is concerned with procedural knowledge and uses skills tests to determine whether students achieve learning objectives. However, the current grading system of "go" and "no go" ratings tends to yield limited variation in training performance, making it difficult to use this outcome for evaluation (e.g., to show whether it is related to any other variables in the model). Thus, an important area for future investigation is to develop and test finer-grained measures of skills that distinguish among levels of performance. Some alternatives include:

- Rate successful completion of different parts of a task, with the final score consisting of the proportion of go ratings
- Measure time to completion in addition to go or no go ratings

- Provide subjective ratings reflecting the quality of performance on procedures using a 1 to 10 scale rather than using go and no go ratings.

To help manage the increased data collection involved with these alternatives, the Army might use hand-held electronic devices to record scores wherever the tests are given and in real time.

Post-training attitudes or affective learning outcomes, such as self-efficacy and learning motivation (Kraiger, Ford, and Salas, 1993) constitute the third measure of "changes in learners." Measures of self-efficacy are strongly related to immediate and delayed procedural knowledge and are particularly valuable when it is impractical to assess learning outcomes (Sitzmann et al., 2008). Consequently, we recommend measuring self-efficacy for procedures taught in DL in order to gauge readiness for resident training (where practice of those procedures typically occurs). Measures of affective learning outcomes can be readily added to end-of-course surveys, although self-efficacy measures are not generic and therefore must be tailored to particular learning objectives for each course. Similar measures administered prior to DL could also be used to examine the extent to which training contributes to changes in affective learning outcomes.

Future studies should also investigate the feasibility of studying transfer performance, e.g., by studying how post-training attitudes, cognitive learning, and training performance influence behavior on the job. As discussed in Chapter Three, a study of the training-performance link would have to go beyond AUTOGEN's current capabilities by distinguishing results from DL and resident phases of a course and by making direct links between students' performance in training and performance on the job. Also, such a study would have to address other challenges, such as the need for measures of performance with sufficient variability to observe associations between scores in training and other outcomes.

Cost-benefit analysis is a critical area for evaluation that falls outside the scope of the integrated training model (which focuses primarily on the benefit side of the equation). TADLP could use such an approach to support requests for DL funding within the programming and budgeting process. Cost-benefit analyses could be used to show, for example, how DL can produce gains such as equivalent learning in reduced training time or cost, or more or better training for the same money and time.

Conduct Evaluation Studies of New DL Approaches

IMI is only one form of DL. In addition to expanding the range of measures used, training evaluation should be extended to other DL approaches in order to determine the measures and methods that are appropriate to evaluate different forms of training, particularly blended learning and mLearning. For example, mLearning involves the use of mobile technologies such as netbooks, tablet computers, electronic book readers, personal digital assistants, and smart phones. Training content presented on mLearning devices is likely to be accessed in small chunks by soldiers on the go. This suggests the need for alternatives to end-of-course surveys or comprehensive tests of learning to evaluate training. For example, small "chunked" evaluations—such as checks on learning and brief reactions measures that are embedded in the delivery platform and interspersed with training content—may be needed to evaluate mLearning outcomes.

Evaluating Changes to Training Policy or Courses

Evaluation is also important to determine the effect of new initiatives within TADLP. In this report we have made a number of recommendations regarding changes to policy, such as arranging training schedules to reduce the time lag between DL and resident training, providing additional administrative support to increase graduation rates, and adding fields to ATRRS to better capture student training status. Another policy change to evaluate in the future is the effect of creating educational duty, or "EDY." Changes to course design to assess include efforts to increase student engagement in IMI courseware or breaking DL products into smaller parts. Collection of baseline training evaluation measures is needed to assess the effects of policy changes or other interventions designed to enhance training quality.

Final Thoughts

Improved tools and metrics for evaluating DL training can provide benefits to TADLP at multiple levels. At the student level, evaluation can enable training staff to determine student success and diagnose needs for remediation. At the course level, evaluation can show how DL affects learning and subsequent outcomes such as knowledge retention and performance on the job; point to needs for improvement in course content or delivery; and determine the effect of interventions designed to enhance training quality or efficiency. At the program level, evaluation can demonstrate the value of DL and support the case for resources to meet program goals.

APPENDIX A

Nongraduate Survey

Army Distributed Learning Survey

The purpose of this survey is to understand why some who enroll in Army distributed learning (DL) courses do not appear to graduate from those courses, and to identify ways to improve the graduation rate. We are a team from the RAND Corporation, a nonprofit research organization, who is working with the Army to conduct and analyze a number of surveys on DL. This survey is focused particularly on those who enrolled in a DL course but did not complete it. The survey will typically take less than 5 minutes to finish. If you did complete the DL course or, alternatively, never enrolled in the course, please indicate that fact in the first question before exiting this site.

The survey asks about your reasons for not completing the DL course, the circumstances under which you were enrolled, and your individual learning preferences.

Your responses to the survey will be anonymous, and your participation in the survey is completely voluntary. RAND will use your responses for research purposes only.

If you have questions about this study, please contact Dr. Susan Straus, 412-683-2300, x4925, sgstraus@rand.org, RAND, 4570 Fifth Avenue, Pittsburgh, PA 15213. If you have questions or concerns about your rights as a research participant, contact James Tebow, Human Subjects Protection Committee, RAND, 1700 Main Street, Santa Monica, CA, 90407, 310-393-0411, x7173. If you need technical support to access or complete the survey, please contact Amy Clark, aclark@rand.org, or 310-393-0411, x6879.

Thank you for contributing to this very important effort for the Army Distributed Learning Program.

You may access this survey using Internet Explorer 6.0+, Firefox, or Safari.

Please click the Next button to begin the survey.

Instructions for Completing the Survey

- For each question, please select a response from the choices provided, or type your response in the text box, as indicated.
- When you are finished with each page, click the Next button at the bottom of the page to advance to go to the next set of questions in the survey.
- To return to a previous screen while taking the survey, click the Back button at the bottom of the page. **Do not** use your browser's navigation buttons or your responses may be lost.

Click Next to go to the first question.

1. Please select the course for which you are taking this survey.

☐ 74D10, CBRN Specialist Course
☐ 88M30, Phase 2 of Motor Transport Operator BNCOC
☐ First Sergeant Course (DL Phase 1)
☐ CMF 63/94 ANCOC (DL Phase)
☐ AMEDD Captains Career Course (DL Phase)

2. Did you complete the DL course named in the email inviting you to take this survey?

☐ Yes
☐ No—I'm currently in the process of taking the DL phase of this course
☐ No—I started taking the DL course but stopped before completing it
☐ No—I enrolled but never started the DL phase of this course
☐ No—I was not aware I was ever enrolled in the DL phase of this course

[IF YES, first NO, or final NO, GO TO END OF SURVEY;
IF second NO—GO TO QUESTION 4 and answer all other questions;
IF third NO—answer questions 3, 5, 6, and 10–16]

We would like to ask you about your enrollment and the circumstances in which you took the course. Please select a response from the choices provided.

3. What were the major contributing reasons you never started the DL course?

Check all that apply.

- ☐ I found that I did not have enough time to work on this course
- ☐ Family or health or work emergency prevented me
- ☐ I was mobilized or deployed
- ☐ I changed occupations, making the course no longer relevant
- ☐ I decided to leave the Army
- ☐ I had difficulty getting access to reliable computer equipment
- ☐ I had difficulty getting access to an Internet connection
- ☐ I had other technical problems with the course (e.g., difficulty launching the courseware)
- ☐ I lost interest
- ☐ Other, please specify ____________________

4. What were the major contributing reasons you *started* taking the DL course, *but never completed it?*

Check all that apply.

- ☐ I found that I did not have enough time to work on this course
- ☐ Family or health or work emergency prevented me
- ☐ I was mobilized or deployed
- ☐ I changed occupations, making the course no longer relevant
- ☐ I decided to leave the Army
- ☐ I had difficulty getting access to reliable computer equipment
- ☐ I had technical problems with the course (e.g., difficulty launching the courseware)
- ☐ I lost interest or felt the course was not worth my time
- ☐ The course was too easy
- ☐ The course was too difficult
- ☐ The course was too long
- ☐ I could not get my questions about the content of the course answered
- ☐ I could not get my questions about the technical aspects of the course answered
- ☐ Other, please specify ____________________

5. When did you enroll in the DL course? Please choose the month in which you enrolled in the DL phase of this course:

- ☐ January
- ☐ February
- ☐ March
- ☐ April
- ☐ May
- ☐ June
- ☐ July
- ☐ August
- ☐ September
- ☐ October
- ☐ November
- ☐ December
- ☐ Do not remember

6. When did you enroll in the DL course? Please choose the year in which you enrolled in the DL course:

- ☐ 2009
- ☐ 2008
- ☐ 2007 or before
- ☐ Do not remember

7. Over what period of time did you work on the DL course?

- ☐ Less than a day
- ☐ More than a day but less than a week
- ☐ More than a week but less than one month
- ☐ Approximately 1–3 months
- ☐ Approximately 4 –6 months
- ☐ Over 6 months

8. Please indicate the type of Internet service you were connected to when you took the course. List the approximate percentage for each option. The sum of the numbers entered must equal 100.

- ☐ High-speed (broadband) connection (for example, DSL, cable, T1 line, or Ethernet connection) (%) ________
- ☐ Dial-Up (%) ________
- ☐ None—I took the course on a CD ROM (%) ________
- ☐ Don't know (%) ________

9. Please indicate your payment status when you took the DL phase of this course. List the approximate percentage for each option. The sum of the numbers entered must equal 100.

- ☐ Paid status or duty hours (%) ________
- ☐ Retirement points only (%) ________
- ☐ Personal time or nonduty hours (%) ________

10. Approximately how many *total hours* of Army DL have you taken in the last year, *excluding* this course?

- ☐ 25 or less
- ☐ 26–50
- ☐ 51–100
- ☐ 101–200
- ☐ More than 200

11. What is your military component?

- ☐ Active component
- ☐ United States Army Reserves (USAR)
- ☐ Army National Guard (ARNG)
- ☐ Military branch other than the Army
- ☐ Civilian

[IF "CIVILIAN"—GO TO QUESTION 14]

12. What is your grade/rank?

- ☐ E-1
- ☐ E-2
- ☐ E-3
- ☐ E-4
- ☐ E-5
- ☐ E-6
- ☐ E-7
- ☐ E-8
- ☐ E-9
- ☐ W-1
- ☐ W-2
- ☐ W-3
- ☐ W-4
- ☐ W-5
- ☐ O-1
- ☐ O-2
- ☐ O-3
- ☐ O-4
- ☐ O-5
- ☐ O-6
- ☐ O-7
- ☐ O-8
- ☐ O-9

[IF (E1–E9)—GO TO QUESTION 13; OTHERWISE, GO TO QUESTION 14]

13. Why did you enroll in this course?

- ☐ The course was required for my job
- ☐ I used the course for self-development
- ☐ I used the course for reachback or refresher training
- ☐ Other (please specify) ____________

14. Please indicate the extent to which you agree or disagree with each statement.

[5-POINT SCALES RANGING FROM STRONGLY DISAGREE TO STONGLY AGREE]

I like to learn:

- ☐ By using a computer (e.g., the Internet or educational software)
- ☐ In a traditional classroom setting
- ☐ By working on my own
- ☐ By working with other students
- ☐ At my own pace
- ☐ At my own convenience, where and when I choose
- ☐ With guidance from an instructor
- ☐ With specific deadlines for assignments

15. The following question asks about <u>information technologies</u>, which include computers, cell phones/smart phones, MP3 players, and other devices. Please indicate the extent to which you agree or disagree with each statement:

[5-POINT SCALES RANGING FROM STRONGLY DISAGREE TO STRONGLY AGREE]

- ☐ I like to experiment with new information technologies
- ☐ Among my peers, I am usually the first to try out new information technologies
- ☐ In general, I am hesitant to try out new information technologies

16. Do you have any additional comments about your enrollment in the DL course that will help us better understand your experience?

[OPEN-ENDED RESPONSE]

Survey Completed

Thank you for taking the survey! You responses have been submitted. We appreciate your time. Your responses are valuable and will help to improve Army distributed learning.

APPENDIX B

Training Circumstances for Nongraduates

Item	Total
Purpose for taking course	
Required	56%
Self-development	43%
Reachback or refresher training	0
Other	<1%
Time to complete course	
Less than a day	7%
More than a day but less than a week	16%
More than a week but less than a month	38%
1–3 months	28%
4–6 months	7%
> 6 months	4%
Medium Used (average %)	
CD-ROM	2%
High-speed internet connection	83%
Dial-up	4%
Don't know	12%
Payment status while taking course (average %)	
Paid/duty hours	31%
Personal time	69%
Retirement points only	<1%

APPENDIX C

Graduate Survey

Distributed Learning Courseware Survey

The purpose of this survey is to understand the quality of Army distributed learning (DL) courses and identify ways to improve training. We are a team from the RAND Corporation, a nonprofit research organization, who is working with the Army to conduct and analyze this survey. The survey will take approximately 15 minutes to complete.

The survey asks about your experience taking the DL phase of a particular training course you have been offered. It includes questions about the circumstances in which you took the course, your individual learning preferences, your overall satisfaction with the course, and the course's content, delivery and technical features.

Your responses to the survey will be anonymous, and your participation in the survey is completely voluntary. RAND will use your responses for research purposes only. We will report to the Army only the average scores in a course; your individual answers will not be provided.

If you have questions about this study, please contact Dr. Susan Straus, 412-683-2300, x4925, sgstraus@rand.org, RAND, 4570 Fifth Avenue, Pittsburgh, PA 15213. If you have questions or concerns about your rights as a research participant, contact James Tebow, Human Subjects Protection Committee, RAND, 1700 Main Street, Santa Monica, CA, 90407, 310-393-0411, x7173. If you need technical support to access or complete the survey, please contact Amy Clark, aclark@rand.org, or 310-393-0411, x6879.

Thank you for contributing to this very important effort for the Army Distributed Learning Program.

You may access this survey using Internet Explorer 6.0+, Firefox, or Safari.

Please click the Next button to begin the survey.

Instructions for Completing the Survey

- For each question, please select a response from the choices provided, or type your response in the text box, as indicated.
- When you are finished with each page, click the Next button at the bottom of the page to advance to the next set of questions in the survey.
- To return to a previous screen while taking the survey, click the Back button at the bottom of the page. **Do not** use your browser's navigation buttons or your responses may be lost.

Click Next to go to the first question.

COURSE SELECTION

1. Please select the course for which you are taking this survey.

- ☐ 21E10
- ☐ 21K10
- ☐ 21R10
- ☐ 21W10
- ☐ 21T10
- ☐ 21N30
- ☐ 21N40
- ☐ 21H30
- ☐ 21H40
- ☐ 31D20/30—Phase I
- ☐ 31D20/30—Phase III
- ☐ 19D
- ☐ 19K
- ☐ 91 A, B, H, M or P
- ☐ AMEDD Captains Career Course (DL Phase)
- ☐ My course is not listed here

2. If you selected "My course is not listed here" above, please enter the course name or number here:

[OPEN-ENDED RESPONSE]

BACKGROUND QUESTIONS

In the first set of questions, we would like to ask you about the circumstances in which you took the course. Please select a response from the choices provided.

3. Did you complete the DL phase of this course?

- ☐ Yes
- ☐ No—I'm currently in the process of taking the DL phase of this course
- ☐ No—I started taking the DL phase of this course but quit before completing it
- ☐ No—I never started the DL phase of this course

[IF first or second NO—GO TO QUESTION 6;
If third NO—GO TO END OF SURVEY]

4. When did you complete the DL phase of this course?
Please choose the month in which you completed the DL phase of this course:

- ☐ January
- ☐ February
- ☐ March
- ☐ April
- ☐ May
- ☐ June
- ☐ July
- ☐ August
- ☐ September
- ☐ October
- ☐ November
- ☐ December

5. Please choose the year in which you completed the DL phase of this course:

- ☐ 2009
- ☐ 2008
- ☐ 2007
- ☐ 2006 or before

6. Over what period of time did you work on the DL phase of this course?

- ☐ Less than one month
- ☐ Approximately 1–3 months
- ☐ Approximately 4 –6 months
- ☐ Over 6 months

7. Was this enough time to work on the course?

- ☐ Yes
- ☐ No
- ☐ N/A—I didn't complete enough of the course to know

8. Please indicate the type of Internet service you were connected to when you took the course. List the approximate percentage for each option. The sum of the numbers entered must equal 100.

- ☐ High-speed (broadband) connection (for example, DSL, cable, T1 line, or Ethernet connection) (%) ____
- ☐ Dial-up (%) ____
- ☐ None—I took the course on a CD ROM (%) ____
- ☐ Don't know (%) ____

9. Where did you complete the majority of the course?

- ☐ At home
- ☐ At a deployed location
- ☐ At a CONUS military facility
- ☐ At civilian office/work/school
- ☐ Other (please specify) ________________

10. Is there a resident phase to this course? ☐ Yes ☐ No [IF "NO"—GO TO QUESTION 12 and SKIP QUESTION 33]
11. When did you complete the DL phase of the course relative to the resident phase? ☐ Before I took any resident phase of the course ☐ At the same time that I was taking a resident phase of the course ☐ After I took a resident phase of the course
12. Please indicate your payment status when you took this DL course. List the approximate percentage for each option. The sum of the numbers entered must equal 100. ☐ Paid status or duty hours (%) ________ ☐ Retirement points only (%) ________ ☐ Personal time or nonduty hours (%) ________
13. Approximately how many total hours of Army DL have you taken in the last year, excluding this course? ☐ 25 or less ☐ 26–50 ☐ 51–100 ☐ 101–200 ☐ More than 200
14. What is your military component? ☐ Active component ☐ United States Army Reserves (USAR) ☐ Army National Guard (ARNG) ☐ Military branch other than the Army ☐ Civilian *[IF "CIVILIAN"—GO TO QUESTION 17]*

15. What is your grade/rank?

- ☐ E-1
- ☐ E-2
- ☐ E-3
- ☐ E-4
- ☐ E-5
- ☐ E-6
- ☐ E-7
- ☐ E-8
- ☐ E-9
- ☐ W-1
- ☐ W-2
- ☐ W-3
- ☐ W-4
- ☐ W-5
- ☐ O-1
- ☐ O-2
- ☐ O-3
- ☐ O-4
- ☐ O-5
- ☐ O-6
- ☐ O-7
- ☐ O-8
- ☐ O-9
- ☐ O-10

[IF (E1–E9)—GO TO QUESTION 16; OTHERWISE, GO TO QUESTION 17]

16. How long have you been serving in the MOS for which you are taking this DL course?

- ☐ I have not served in this MOS
- ☐ Less than 6 months
- ☐ 6 months–11 months
- ☐ 1–2 years
- ☐ Over 2 years
- ☐ This course is not specific to my MOS

17. How much civilian experience do you have related to the subject matter of this course?

- ☐ None
- ☐ A small amount
- ☐ A moderate amount
- ☐ A substantial amount

18. Why did you take this course?

- ☐ The course was required for my job
- ☐ I used the course for self-development
- ☐ I used the course for reachback or refresher training
- ☐ Other (please specify) ____________

TECHNICAL FEATURES, USER INTERFACE AND SUPPORT

The next section asks about the user interface and other technical aspects, and the support received while taking this DL course. Please select a response from the choices provided.

19. Did you experience any of the following in this course?

[YES; NO; DON'T KNOW/ DON'T REMEMBER; NOT APPLICABLE FOR EACH]

- ☐ Difficulty registering for the course
- ☐ Difficulty accessing the courseware over the Internet
- ☐ Difficulty receiving the courseware on a CD in the mail
- ☐ Difficulty launching the courseware
- ☐ Difficulty navigating through the courseware
- ☐ Difficulty determining where you were within the course
- ☐ Difficulty playing audio, video, or running animations in the courseware
- ☐ Delay in pages loading
- ☐ Difficulty returning to the spot where you left off after logging off or closing the courseware, intentionally or unintentionally
- ☐ Lost data such as scores on practical exercises or tests
- ☐ Text on the screen that was hard to read
- ☐ Audio narration that was too slow or too fast
- ☐ Sounds, graphics, or animations that were distracting

20. Did you experience any other technical difficulties during this course that were not mentioned in the previous question? If so, please describe in the space provided below:

[OPEN-ENDED RESPONSE]

21. Overall, how satisfied were you with your experience in using the course website as well as technical support you received while taking this course?

- ☐ Very Dissatisfied
- ☐ Dissatisfied
- ☐ Neutral
- ☐ Satisfied
- ☐ Very Satisfied

22. Did you require technical support while taking the DL phase of this course?

- ☐ Yes
- ☐ No

[IF "NO"—GO TO QUESTION 25]

23. Please indicate which types of support you used to get help with technical issues, and how satisfied you were with each type of support you used. For technical issues, did you:

[FOR EACH ITEM, OPTIONS ARE: NO; YES—VERY DISSATISFIED; YES—DISSATISFIED; YES—NEUTRAL, YES—SATISFIED; YES—VERY SATISFIED]

- ☐ Contact the Army help desk
- ☐ Contact the help desk or other support staff at the proponent school
- ☐ Ask an instructor
- ☐ Other

24. If you indicated "Yes" for "Other" in the previous question, please describe the type of support you used in the space provided below:

[OPEN-ENDED RESPONSE]

25. Did this course include supporting materials, such as field manuals and glossaries?

- ☐ Yes
- ☐ No
- ☐ Don't Know

[IF "NO"—GO TO QUESTION 27]

26. The supporting materials such as field manuals and glossaries were useful. ☐ Strongly Disagree ☐ Disagree ☐ Neutral ☐ Agree ☐ Strongly Agree ☐ N/A—I didn't use supporting materials
27. If I had a question about the course content, I was able to get an answer or explanation quickly and completely from either the instructor or supporting materials. ☐ Strongly Disagree ☐ Disagree ☐ Neutral ☐ Agree ☐ Strongly Agree ☐ N/A—I didn't have any questions about course content
28. Do you have any comments about the user interface, technical aspects, or support for the course? *[OPEN-ENDED RESPONSE]*
COURSE CONTENT AND DELIVERY *The next set of questions asks about your views of the course content and delivery, including how material was presented, the relevance of the content,, the quality of opportunities for practice, how engaging the course was, and interaction with instructors and other students.*
29. The course clearly explained important terms and concepts. ☐ Strongly Disagree ☐ Disagree ☐ Neutral ☐ Agree ☐ Strongly Agree

30. The course clearly demonstrated how to perform procedures. ☐ Strongly Disagree ☐ Disagree ☐ Neutral ☐ Agree ☐ Strongly Agree ☐ N/A—The course did not include procedures
31. It was clear how the course material applies to my job. ☐ Strongly Disagree ☐ Disagree ☐ Neutral ☐ Agree ☐ Strongly Agree ☐ N/A—The course was not specific to my job
32. As a result of taking this course, I will be able to perform better in my job or during an operation. ☐ Strongly Disagree ☐ Disagree ☐ Neutral ☐ Agree ☐ Strongly Agree ☐ N/A—The course was not specific to my job
33. The course helped me prepare for the resident phase of training. ☐ Strongly Disagree ☐ Disagree ☐ Neutral ☐ Agree ☐ Strongly Agree ☐ N/A—I have not started or do not plan to take the resident phase of this course

34. Please indicate the extent to which you agree or disagree with each statement:

[5-POINT SCALES RANGING FROM STRONGLY DISAGREE TO STRONGLY AGREE.]

- ☐ Practical exercises and checks on learning helped me learn the course material*
- ☐ There were enough practical exercises and checks on learning in the course
- ☐ If I made a mistake, the feedback adequately explained why I was wrong
- ☐ The course held my interest
- ☐ I clicked through a lot of the screens without really paying attention
- ☐ I already knew a lot of the material covered in this course

** Checks on learning consist of brief questions throughout the lessons to test your understanding of the material. Practical exercises are usually found at the end of a lesson and require hands-on practice or application of the material.*

35. The level of difficulty of the course was about right.

- ☐ Strongly Disagree
- ☐ Disagree
- ☐ Neutral
- ☐ Agree
- ☐ Strongly Agree

[IF NEUTRAL, AGREE OR STRONGLY AGREE, GO TO QUESTION 37]

36. The level of difficulty of the course was:

- ☐ Too easy
- ☐ Too hard

37. The length of the course (amount of course material) was about right.

- ☐ Strongly Disagree
- ☐ Disagree
- ☐ Neutral
- ☐ Agree
- ☐ Strongly Agree

[IF NEUTRAL, AGREE OR STRONGLY AGREE, GO TO QUESTION 39]

38. The length of the course was:

- ☐ Too short
- ☐ Too long

39. Did you interact with an instructor about course content?

- ☐ Yes
- ☐ No

[IF "NO"—GO TO QUESTION 41]

40. I was satisfied with the level of interaction with the instructor.

- ☐ Strongly Disagree
- ☐ Disagree
- ☐ Neutral
- ☐ Agree
- ☐ Strongly Agree

[5-POINT SCALE RANGING FROM STRONGLY DISAGREE TO STRONGLY AGREE; IF NEUTRAL, AGREE OR STRONGLY AGREE—GO TO QUESTION 42]

41. The amount of interaction with the instructor was:

- ☐ Too little
- ☐ Too much

42. Did you interact with other students about course content?

- ☐ Yes
- ☐ No

[IF "NO"—GO TO QUESTION 44]

43. I was satisfied with the level of interaction with other students.

- ☐ Strongly Disagree
- ☐ Disagree
- ☐ Neutral
- ☐ Agree
- ☐ Strongly Agree

[IF NEUTRAL, AGREE OR STRONGLY AGREE—GO TO QUESTION 45]

44. The amount of interaction with the other students was:

- ☐ Too little
- ☐ Too much

45. Do you have any comments about the course content, practical exercises and checks on learning, how engaging the course was, or interaction with instructors or students?

[OPEN-ENDED RESPONSE]

OVERALL SATISFACTION

The next set of questions asks about your overall satisfaction with the course.

46. Please indicate the extent to which you agree or disagree with each statement:

[5-POINT SCALES RANGING FROM STRONGLY DISAGREE TO STRONGLY AGREE.]

- ☐ Overall, I was satisfied with the DL phase of this course
- ☐ I look forward to taking another DL course
- ☐ The DL phase of this course was a good use of my time
- ☐ Based on my experience in this course, I would recommend DL courses to others

47. Overall, which of the following had the biggest effect on your satisfaction with this course?

[PLEASE RANK FACTORS FROM 1 TO 4 WITH 1 = FACTOR HAVING THE BIGGEST EFFECT AND 4 = FACTOR HAVING THE SMALLEST EFFECT.]

- ☐ The technical features of the course, including courseware/technical support ____
- ☐ Course content ____
- ☐ The amount of time I had to work on the course ____
- ☐ Other ____

48. If you assigned a numerical rank for "Other" in the previous question, please describe the factor affecting your satisfaction in the space provided below:

[OPEN-ENDED RESPONSE]

INDIVIDUAL PREFERENCES

The last set of questions asks about the ways in which you prefer to learn.

49. Please indicate the extent to which you agree or disagree with each statement.

[5-POINT SCALE FOR EACH, RANGING FROM STRONGLY DISAGREE TO STONGLY AGREE.]

I like to learn:

- ☐ By using a computer (e.g., the Internet or educational software)
- ☐ In a traditional classroom setting
- ☐ By working on my own
- ☐ By working with other students
- ☐ At my own pace
- ☐ At my own convenience, where and when I choose
- ☐ With guidance from an instructor
- ☐ With specific deadlines for assignments

50. The following question asks about information technologies, which include computers, cell phones/smart phones, MP3 players, and other devices. Please indicate the extent to which you agree or disagree with each statement: *[5-POINT SCALES RANGING FROM STRONGLY DISAGREE TO STRONGLY AGREE.]* ☐ I like to experiment with new information technologies ☐ Among my peers, I am usually the first to try out new information technologies ☐ In general, I am hesitant to try out new information technologies
51. Do you have any additional comments about the DL phase of this course that will help us better understand your experience? *[OPEN-ENDED RESPONSE]*
Thank you for taking the survey! You responses have been submitted. We appreciate your time. Your responses are valuable and will help to improve Army distributed learning.

APPENDIX D

Graduate Survey Participant Characteristics

Item	Percent
Military Component	
AC	63%
USAR	14%
ARNG	23%
Grade or Rank	
E-4	4%
E-5	4%
E-6	19%
E-7	19%
E-8	6%
O-2	6%
O-3	33%
O-4	7%
Other	<3%
Prior experience in MOS	
None	18%
< 6 months	11%
6–11 months	18%
1–2 years	10%
> 2 years	18%
N/A	25%
Prior civilian experience	
None	67%
Small amount	20%
Moderate amount	9%
Substantial amount	4%

APPENDIX E

DL Training Circumstances for Graduates

Item	Total
Purpose of taking course	
Required	77%
Self-development	23%
Time to complete course	
< 1 month	52%
1–3 months	35%
4–6 months	7%
> 6 months	6%
Location (%)	
Home	42%
Deployed	13%
CONUS military facility	35%
Civilian office or school	10%
Combination or other	<1%
Percentage of each medium used while working on course	
CD-ROM	12%
High-speed Internet Connection (%)	77%
Dial-Up (%)	2%
Don't know (%)	7%
Average time spent taking course across type of payment status	
Paid/duty hours	47%
Personal time	53%
Retirement points only	< 1%
When DL was completed	
Before resident phase	90%
At the same time as resident phase	9%
After resident phase	1%

APPENDIX F

Revised Nongraduate Survey

Army Distributed Learning Survey

The purpose of this survey is to understand why some who enroll in Army distributed learning (DL) courses do not appear to graduate from those courses, and to identify ways to improve the graduation rate. The survey will take less than 5 minutes to finish. If you did complete the DL course or, alternatively, never enrolled in the course, please indicate that fact in the first question before exiting this site.

The survey asks about your reasons for not completing the DL course and the circumstances under which you were enrolled. Your responses to the survey will be anonymous.

Thank you for contributing to this very important effort for the Army Distributed Learning Program.

Please click the Next button to begin the survey.

Instructions for Completing the Survey

- For each question, please select a response from the choices provided, or type your response in the text box, as indicated.
- When you are finished with each page, click the Next button at the bottom of the page to advance to go to the next set of questions in the survey.
- To return to a previous screen while taking the survey, click the Back button at the bottom of the page. Do not use your browser's navigation buttons or your responses may be lost.

Click Next to go to the first question.

1. Please select the course for which you are taking this survey:
[Customize options for the course.]

2. Did you complete the DL course named in the message asking you to complete this survey?

a. Yes
b. No—I'm currently in the process of taking the DL phase of this course
c. No—I started taking the DL course but stopped before completing it
d. No—I enrolled but never started the DL phase of this course
e. No—I was not aware I was ever enrolled in the DL phase of this course
f. No—I was enrolled in error

[IF YES, answer questions 10–12
IF first NO, GO TO END OF SURVEY;
IF second NO—GO TO QUESTION 3 and answer all other questions;
IF third NO—answer questions 2, 4, 5, and 9–13;
IF fourth NO—answer questions 10–11;
IF final NO—answer questions 10–11]

The next set of questions asks about your enrollment and the circumstances in which you took the course.

3. What were the major contributing reasons you never started the DL course? Check all that apply. a. I found that I did not have enough time to work on this course b. Family or health or work emergency prevented me c. I was mobilized or deployed d. I changed occupations, making the course no longer relevant e. I decided to leave the Army f. I lost interest or felt that the course was not worth my time g. I had difficulty getting access to reliable computer equipment h. I had difficulty getting access to an Internet connection i. I had other technical problems with the course (e.g., difficulty launching the courseware) (Please specify__________________). j. I could not get my questions about the technical aspects of the course answered k. Other, please specify ____________________
4. What were the major contributing reasons you started taking the DL course, *but never completed it?* Check all that apply. a. I found that I did not have enough time to work on this course b. Family or health or work emergency prevented me c. I was mobilized or deployed d. I changed occupations, making the course no longer relevant e. I decided to leave the Army f. I had difficulty getting access to reliable computer equipment g. I had difficulty getting access to an Internet connection h. I had technical problems with the course (e.g., difficulty launching the courseware) i. I lost interest or felt the course was not worth my time j. The course was too easy k. The course was too difficult l. The course was too long m. I could not get my questions about the content of the course answered n. I could not get my questions about the technical aspects of the course answered o. Other, please specify ____________________

5. In what month did you enroll in the DL phase of this course?

a. January
b. February
c. March
d. April
e. May
f. June
g. July
h. August
i. September
j. October
k. November
l. December
m. Do not remember

6. In what year did you enroll in the DL phase of this course?

a. 2009
b. 2008
c. 2007 or before
d. Do not remember

7. Over what period of time did you work on the DL course?

a. Less than a day
b. More than a day but less than a week
c. More than a week but less than one month
d. Approximately 1–3 months
e. Approximately 4 –6 months
f. Over 6 months

8. What type of Internet service did you use to take all or the majority of the course?

a. High-speed (broadband) connection (for example, DSL, cable, T1 line, or Ethernet connection)
b. Dial-up
c. None—I took the course on CD-ROM
d. Don't know

9. Please indicate your payment status (or your expected payment status) when you took (or were planning to take) the DL phase of this course. List the approximate percentage for each option. The sum of the numbers entered must equal 100. a. Paid status or duty hours (%) ________ b. Retirement points only (%) ________ c. Personal time or nonduty hours (%) ________ d. Unknown (%)________
10. What is your military component? a. Active component b. United States Army Reserves (USAR) c. Army National Guard (ARNG) d. Military branch other than the Army e. Civilian *[IF "CIVILIAN"—GO TO QUESTION 12]*
11. What is your grade/rank? [customize response options to the course]
12. Why did you enroll in this course? a. Graduation from the course was required for my present job or for a future job b. I took the course for self-development and was seeking to graduate c. I enrolled for self-development, reachback, or refresher training and did not seek graduation d. Other (please specify) ____________
13. Do you have any additional comments about your enrollment in the DL course that will help us better understand your experience? *[OPEN-ENDED RESPONSE]*
Thank you for completing the survey! You responses have been submitted. We appreciate your time. Your responses are valuable and will help to improve the Army distributed learning program.

APPENDIX G

Revised Graduate Survey

Distributed Learning Courseware Survey

The purpose of this survey is to understand the quality of Army distributed learning (DL) courses and identify ways to improve training. The survey asks about your experience taking the DL phase of a particular training course. It includes questions about the circumstances in which you took the course, your overall satisfaction with the course, and the course's content, delivery and technical features.

Your responses to the survey will be anonymous

Thank you for contributing to this very important effort for the Army Distributed Learning Program.

Please click the Next button to begin the survey.

Instructions for Completing the Survey

- For each question, please select a response from the choices provided, or type your response in the text box, as indicated.
- When you are finished with each page, click the Next button at the bottom of the page to advance to the next set of questions in the survey.
- To return to a previous screen while taking the survey, click the Back button at the bottom of the page. Do not use your browser's navigation buttons or your responses may be lost.

Click Next to go to the first question.

COURSE SELECTION

1. Please select the course for which you are taking this survey. [Customize options for the course.]

2. Did you complete the DL phase of this course?

a. Yes
b. No—I'm currently in the process of taking the DL phase of this course

[IF NO—GO TO END OF SURVEY]

BACKGROUND QUESTIONS

In the next set of questions, we would like to ask you about the circumstances in which you took the course. Please select a response from the choices provided.

3. Over what period of time did you work on the DL phase of this course?

a. Less than one month
b. Approximately 1–3 months
c. Approximately 4 –6 months
d. Over 6 months

4. What type of Internet service did you use to take all or the majority of the course?

a. High-speed (broadband) connection (for example, DSL, cable, T1 line, or Ethernet connection)
b. Dial-up
c. None—I took the course on CD-ROM
d. Don't know

5. Where did you complete the majority of the course?

a. At home
b. At a deployed location
c. At a CONUS military facility
d. At civilian office/work/school
e. Other (please specify) ______________

6. Please indicate your payment status when you took this DL course. List the approximate percentage for each option. The sum of the numbers entered must equal 100. a. Paid status or duty hours (%) ________ b. Retirement points only (%) ________ c. Personal time or nonduty hours (%) ________
7. What is your military component? a. Active component b. United States Army Reserves (USAR) c. Army National Guard (ARNG) d. Military branch other than the Army e. Civilian *[IF "CIVILIAN"—GO TO QUESTION 9]*
8. What is your grade/rank? [customize response options to the course]
9. How long have you been serving in the MOS for which you are taking this DL course? a. I have not served in this MOS b. Less than 6 months c. 6 months–11 months d. 1–2 years e. Over 2 years f. N/A or this course is not specific to my MOS
10. How much civilian experience do you have related to the subject matter of this course? a. None b. A small amount c. A moderate amount d. A substantial amount

11. Why did you take this course?

a. The course was required for my job
b. I used the course for self-development
c. I used the course for reachback or refresher training
d. Other (please specify) ___________

TECHNICAL FEATURES, USER INTERFACE AND SUPPORT

The next section asks about the user interface and other technical aspects, and the support received while taking this DL course. Please select a response from the choices provided.

12. Did you experience any of the following in this course?

[RESPONSE OPTIONS FOR EACH ITEM ARE YES; NO; DON'T KNOW/DON'T REMEMBER; NOT APPLICABLE]

a. Difficulty registering for the course
b. Difficulty accessing the courseware over the Internet
c. Difficulty receiving the courseware on a CD in the mail
d. Difficulty launching the courseware
e. Difficulty navigating through the courseware
f. Difficulty determining where you were within the course
g. Difficulty playing audio, video, or running animations in the courseware
h. Delays in pages loading
i. Difficulty returning to the spot where you left off after logging off or closing the courseware, intentionally or unintentionally
j. Lost data such as scores on practical exercises or tests
k. Text on the screen that was hard to read
l. Audio narration that was too slow or too fast
m. Distracting sounds, graphics, or animations
n. Other: (please describe)

13. Overall, how satisfied were you with the technical features of the course?

a. Very Dissatisfied
b. Dissatisfied
c. Neutral
d. Satisfied
e. Very Satisfied

14. Please indicate which types of support you used to get help with technical issues, and how satisfied you were with each type of support you used. For technical issues, did you:

[FOR ITEMS a–d, OPTIONS ARE: NO; YES—VERY DISSATISFIED; YES—DISSATISFIED; YES—NEUTRAL, YES—SATISFIED; YES—VERY SATISFIED]

a. Contact the Army help desk
b. Contact the help desk or other support staff at the proponent school
c. Ask an instructor
d. Other
e. I did not need technical support
f. I needed technical support but did not seek it

15. Supporting materials such as field manuals and glossaries were useful.

a. Strongly Disagree
b. Disagree
c. Neutral
d. Agree
e. Strongly Agree
f. N/A—I didn't use supporting materials
g. N/A—My course did not have supporting materials

16. If I had a question about the course content, I was able to get an answer or explanation quickly and completely from either the instructor or supporting materials.

a. Strongly Disagree
b. Disagree
c. Neutral
d. Agree
e. Strongly Agree
f. N/A—I didn't have any questions about course content

COURSE CONTENT AND DELIVERY

The next set of questions asks about your views of the course content and delivery, including how material was presented, the relevance of the content, the quality of opportunities for practice, how engaging the course was, and interaction with instructors and other students.

17. The course clearly explained important terms and concepts.

a. Strongly Disagree
b. Disagree
c. Neutral
d. Agree
e. Strongly Agree

18. The course clearly demonstrated how to perform procedures.

a. Strongly Disagree
b. Disagree
c. Neutral
d. Agree
e. Strongly Agree
f. N/A—The course did not include procedures

19. It was clear how the course material applies to my job.

a. Strongly Disagree
b. Disagree
c. Neutral
d. Agree
e. Strongly Agree
f. N/A—The course was not specific to my job

20. As a result of taking this course, I will be able to perform better in my job or during an operation.

a. Strongly Disagree
b. Disagree
c. Neutral
d. Agree
e. Strongly Agree
f. N/A—The course was not specific to my job

21. The course helped me prepare for the resident phase of training.

a. Strongly Disagree
b. Disagree
c. Neutral
d. Agree
e. Strongly Agree
f. N/A—I do not plan to take the resident phase of this course

22. Practical exercises and checks on learning helped me learn the course material*

a. Strongly Disagree
b. Disagree
c. Neutral
d. Agree
e. Strongly Agree

* *Checks on learning consist of brief questions throughout the lessons to test your understanding of the material. Practical exercises are usually found at the end of a lesson and require hands-on practice or application of the material.*

23. There were enough practical exercises and checks on learning in the course

a. Strongly Disagree
b. Disagree
c. Neutral
d. Agree
e. Strongly Agree

24. If I made a mistake, the feedback adequately explained why I was wrong

a. Strongly Disagree
b. Disagree
c. Neutral
d. Agree
e. Strongly Agree

25. The course held my interest

a. Strongly Disagree
b. Disagree
c. Neutral
d. Agree
e. Strongly Agree

26. I clicked through a lot of the screens without really paying attention

a. Strongly Disagree
b. Disagree
c. Neutral
d. Agree
e. Strongly Agree

27. The level of difficulty of the course was:

a. Too easy
b. About right
c. Too hard

28. The length of the course was

a. Too short
b. About right
c. Too long

29. The amount of interaction with the instructor was:

a. Too little
b. About right
c. Too much

30. The amount of interaction with the other students was:

a. Too little
b. About right
c. Too much

OVERALL SATISFACTION

The next set of questions asks about your overall satisfaction with the course.

Please indicate the extent to which you agree or disagree with each statement:

[5-POINT SCALES RANGING FROM STRONGLY DISAGREE TO STRONGLY AGREE.]

31. Overall, I was satisfied with the DL phase of this course

   a. Strongly Disagree
   b. Disagree
   c. Neutral
   d. Agree
   e. Strongly Agree

32. I look forward to taking another DL course

   a. Strongly Disagree
   b. Disagree
   c. Neutral
   d. Agree
   e. Strongly Agree

33. The DL phase of this course was a good use of my time

   a. Strongly Disagree
   b. Disagree
   c. Neutral
   d. Agree
   e. Strongly Agree

34. Based on my experience in this course, I would recommend DL courses to others

   a. Strongly Disagree
   b. Disagree
   c. Neutral
   d. Agree
   e. Strongly Agree

35. Do you have any additional comments about the DL phase of this course that will help us better understand your experience? *[OPEN-ENDED RESPONSE]*
Thank you for completing the survey! You responses have been submitted. We appreciate your time. Your responses are valuable and will help to improve the Army distributed learning program.

APPENDIX H

Scoring Procedures for Student Surveys

Nongraduate Survey

Table H.1 shows how the options in Questions 3 and 4 are grouped into categories of reasons for nongraduation. Frequencies for each category are calculated based on whether a student checks one or more options within the category. For example, a student who selects both "3c" and "3d" would have a frequency of "1" for "External Factors," as would a student who selected only "3c" or "3d." However, students can be counted in more than one category; for example, they could respond that they were deployed (an external factor) and also that they could not get access to a reliable computer (a technical factor).

Table H.1
Scoring Procedures for Nongraduate Survey

	Items	
Category	**Enrolled in Course but Did Not Start (Question 3)**	**Started Course but Did Not Complete It (Question 4)**
External factors	b, c, d, e	b, c, d, e
DL factors		
Technical problems	g, h, i	f, g, h
Support	j	m, n
Time	a	a
Courseware	f	i, j, k, l

Graduate Survey

Responses to items with 5-point response options (strongly disagree to strongly agree) are scored so that strongly disagree = 1, disagree = 2, neutral = 3, agree = 4, and strongly agree = 5. These scores will be averaged, as described below. Some of these items also have options such as "not applicable." These responses do not get converted to a number; instead, calculate the percentage of students who selected these options.

For constructs with multiple items and an "Average" scoring method (such as learner engagement) in Table H.2, calculate each student's average across the items. Using these numbers, calculate the average rating across all students. For example, a student who gave ratings of 3 and 4 to items 25 and 26, respectively, would have a learner engagement score of 3.5 ((3 + 4)/2). This score would be averaged with the learner engagement scores for other students. Constructs consisting of only one item with an "Average" scoring method (such as "Overall Satisfaction with Technical Features") may be averaged across students as is.

For constructs that use the "Frequency" scoring method, calculate the number and percentage of responses for each response option across all students. For example, assume 50 students answer the item about course length. If 5 responded that the course was too short, 32 responded that the course was just right, and 13 responded that it was too long, the percentages would be 10 percent, 64 percent, and 26 percent, respectively.

For items that use "Frequency" scoring, if only a small percentage of students select a particular option, it may be worthwhile to combine it with another option. For example, if only 2 percent of students report having "a substantial amount" of civilian experience in response to Question 10, it may be appropriate to combine these results with students who selected "a moderate amount."

Table H.2
Scoring Procedures for Graduate Survey

Construct	Items	Scoring Method
Background questions	1–11	Frequency
Technical problems	12a–n	Frequency Group responses as follows: Access: a and b or c Bandwidth or speed: g,h, and j Navigation: e, f, and i Production quality: k, l, and m Other: n
Overall satisfaction with technical features	13	Average
Technical support	14	Average score for each option (a–d) for which students responded "Yes." Calculate frequencies for options e and f.
Support for content—materials	15	Average
Support for content—overall	16	Average
Course content and delivery	17–24	Average
Learner engagement	25–26	Average; reverse score Item 26 [a]
Course difficulty	27	Frequency
Course length	28	Frequency
Interaction with instructors	29	Frequency
Interaction with students	30	Frequency
Overall satisfaction	31–34	Average

[a] Item 26 is negatively worded such that higher ratings indicate a less favorable response. Therefore, it must be reverse scored before averaging responses with Item 25. To reverse score the item, recode ratings of 1 to 5, 2 to 4, 4 to 2, and 5 to 1.

APPENDIX I

Questions for Semi-Structured Interviews with SMEs About Army Information Systems

General Questions

A. Your role

1. What is your role in designing, developing, maintaining, managing, using or interacting with IT systems that support Army training and leader development?

B. Automatic data collection

1. Would it be valuable to DA, TRADOC, or others to measure, collect, and share better data on student performance, and if so, how might such data be collected?
2. What might be the value to TRADOC or others of automatically collecting web-product usage via web-logs in order to determine such user behaviors as linger-time on each web page, looping, or revisiting of pages?
3. Would it be useful to provide online course developers with a "dashboard" to show them usage and evaluation of their courses in real time?

C. Student or expert course evaluation

1. Would it be valuable to use after-course surveys of Web-based courses to allow students to evaluate course content, their

overall experience in taking the course, or detailed page-by-page content and behavior of the Web course?

2. Would nonstudent subject-matter expert evaluations be valuable as well?

3. Are there evaluation data that are currently generated but not collected and shared, for example, student surveys after completing blended learning courses, or after action reviews involving instructors and students?

D. Course development and delivery tool instrumentation

1. Would it be useful to instrument course development and delivery tools to attempt to measure their utility for course developers?

2. What aspects of a course and its context would it be useful to include for this purpose, e.g., class size, frequency, delivery methods, complexity of interaction (for IMI), in-house versus outsourced development modes, etc?

Questions About Specific Systems You Use

E. Data currently being collected and analyzed

1. Does the system currently collect data regarding student performance in courses, course completion rates or student experiences? If so what is collected? If not, can analysts get such data from other systems, databases, or other sources?

2. What kinds of data about student performance on tests (at the test level and item level) can course managers get and analyze from the system?

3. What kinds of analysis (if any) are performed on system, course, student performance, or training outcome data?

4. Does the system allow students to reach back to review course content or their own performance after they have graduated from a course?

5. How widely is the system used across the schools?

F. Technical

1. What kinds of evaluation and/or reporting capabilities are currently in the system, to what extent are these actually used, and to what extent are they automatic?

2. What kinds of additional evaluation and/or reporting capabilities have been considered for being added to the system, what evidence is there that these would improve training, and what would be involved in adding and using them?

3. What kinds of additional evaluation and/or reporting capabilities could conceivably be added to the system, what evidence is there that these would improve training, and what would be involved in adding and using them?

4. How (if at all) does the system currently interact and interoperate with other ATLD systems? Would additional interaction be necessary or useful to enable evaluation and reporting? If so, how much retrofitting, reimplementation, redesign, or re-architecting would be required to do this?

5. Are evaluation data standardized across Army systems, and if not, should they be, and what would in involved in doing so?

G. Methodological

1. To the extent that the system currently supports evaluation or is envisioned as doing so in the future, what kinds of analysis does it support? For example, does it measure and report individual test item responses, response times, linger time, etc.?

2. How does the system deal with population issues, such as variations among students with different backgrounds or specialties?

3. Does the system collect information automatically within tests, by survey, or by other means?

4. Are tests and evaluation data valid, reliable, and comparable across Army training systems, and if not, what would be involved in making them so?

H. Organizational/cultural issues surrounding collecting, sharing and analyzing course, student, or training evaluation data

1. Are there current policies that would enable or impede the collection, sharing or analysis of such data? If not, should there be such policies?

2. What policies or other mechanisms would best enable the collection, sharing or analysis of such data, without making it simply another compliance issue?

3. What organizational issues or current practices affect the collection, sharing, and analysis of evaluation data by the system? For example, would organizational boundaries between individual schools, TRADOC HQ, and DA make sharing and analysis difficult? Are there traditions of keeping evaluation data private within the organization that manages the system or within the schools that run it?

4. Are there privacy concerns about collecting or sharing individual-level evaluation data? If so, could de-identification or aggregation of such data address these concerns?

5. Do you have staff with appropriate skills for analyzing such data and performing such evaluations? If not, what would be involved in obtaining such staff or contracting out such activities?

I. LMS-specific questions

1. Do you use ALMS, BlackBoard (BB), or a different LMS?

2. If so, did the transition involve any problems in terms of implementing or re-implementing course or student evaluation? Have you had any other problems using ALMS?

3. Is the lack of instructor-in-the-loop interaction in ALMS a limitation in terms of being able to evaluate students?

APPENDIX J

Service-Oriented Architecture

Background

Service-oriented architecture (SOA) is a computational architecture: that is, it is a way of designing, organizing, and invoking computer programs to perform a range of processes. Examples of other (previous) computational architectures include batch processing, timesharing on a mainframe computer, standalone personal computers (PCs), networked workstations, client-server, three-tiered, and multi-tiered architectures. SOA is not a network architecture, though it has some implications for network design and places some demands on network performance. It is a software architecture only in the broad sense that it argues for dividing software up into remotely invokable "services," each of which provides a specific function that is of fairly general utility.

SOA is a general approach to creating and organizing computational capabilities and making them available to users. In particular, SOA centers around the notion of services. A service is a program that runs on some computer (a server) that is attached to a network, enabling the program to be invoked over that network to perform some computation. Services are therefore a particular kind of software system, intended to perform separable computations that can be combined with those of other services to perform a wide range of business processes, which are not constrained or even known in advance. A simple analogy is a set of construction services, such as plumbers, carpenters, electricians, roofers, painters, etc., which can be combined, organized, sequenced, scheduled, and enlisted by a contractor to build a house, remodel or upgrade an existing house, construct a garage or pool, etc.

SOA is the latest incarnation of a long-standing goal among computer scientists to achieve a degree of compatibility, interoperability, interchangeability, and reuse among coarse-grained software components that approaches that of the hardware components that are routinely used in many branches of engineering, including computer hardware design. Throughout its relatively short history, software has suffered from a notorious lack of these qualities. Working computer programs must often be adapted, rebuilt, or redesigned in order to be used in new systems or environments or for new purposes, leading to low levels of reuse and high cost for upgrading or replacing existing IT systems. Unlike hardware components, whose precise specifications are available to engineers in catalogues and databases, software components have traditionally been difficult to describe, identify, or locate and difficult to adapt and integrate into new systems even when they exist and can be found. Numerous concepts such as that of a "software bus" (or "software backplane"), software component warehouses or repositories, mega-programming, and "plug-and-play" components have been proposed to try to address this issue, but none of them have had resounding success to date.

SOA is an attempt to leverage the ubiquity of the Internet (and in particular the World Wide Web protocols that run on top of the Internet) to enable runtime "services" to be engaged in a modular fashion to perform larger-scale computations. Although SOA can be seen as an evolution of client-server and three-tiered (or multi-tiered) architectures, it differs from these in a number of important ways. There are several distinct forms of SOA, but they all share several core attributes:

- A networked environment that enables "services" to run on distributed host computers.
- Facilities to enable services to publish their existence, their capabilities, and their interfaces so that users (including other services) can discover them at runtime and determine their appropriateness and utility for a desired purpose.
- Facilities to enable services to invoke each other to perform various computations, communicating input data and results over the network.

- Facilities to enable users to "orchestrate" the invocation of appropriate sequences of services to perform desired business processes.

The grand vision of SOA is to provide an open marketplace in which services can be created and published by any organization, vendor, or individual and can then be discovered and invoked at runtime by any other service or user on the Internet, without prior arrangement with the service's creator.

In order to realize its interoperability goals, the most popular forms of SOA rely on a number of formalized specification languages and protocols to allow services to describe their capabilities and interfaces, specify their security and authentication policies, and publish the information needed to invoke them. These formalizations are typically encoded in XML, making them relatively easy for humans to understand, as well as being interpretable by computer programs. The resulting specifications are intended to enable services to interoperate with each other whether or not they were designed together or with any knowledge of each other's existence. The process of invoking a sequence of services to perform a particular business process is referred to as "orchestration" and is performed using one of another set of specifications (essentially simplified programming languages) that are also defined as part of the SOA environment.

Unlike many previous forms of software componentization, SOA relies on running services that can be discovered and invoked at runtime, as opposed to static code modules that must be assembled into a program or system and then run. Aside from this key distinction, however, SOA can be seen as yet another form of modularization, i.e., a method for allowing designers and implementers to divide large-scale problems into small-scale pieces, many of which will hopefully already exist. This "divide-and-conquer" approach is fundamental to solving any complex problem and is reflected in the many modularization mechanisms that have been developed by computer science, such as subroutines, macros, data types, objects, aspects, and virtual machines.

The underlying challenge of any divide-and-conquer approach is how to divide a complex problem into appropriate pieces, often referred to as factorization. Different modularization approaches may favor different factorization strategies, but it is a truism that no single factorization is optimal (or even sufficient) for all problems. As a simple analogy, printing a list of people sorted by their names makes it easy to find an individual by name but difficult to find one by address or phone number. More to the point, defining SOA services inappropriately may result in excessive amounts of network traffic to pass input data and results among the multiple services needed to perform a given task.

General criteria for successful factorization include minimizing "coupling" among distinct modules while maximizing "cohesion" within each module; but these attributes of a given factorization tend to be problem-specific, making it difficult to create a set of services that can be combined to solve a wide range of problems while minimizing their coupling and maximizing their cohesion in all cases. Moreover, no matter how a problem is factored, many "cross-cutting concerns" such as performance, usability, availability, reliability, security, extensibility, and interoperability are pervasive and fall across all modules, making any factorization equally poor at addressing such concerns. This fundamental factorization challenge is inherent in every divide-and-conquer approach that has yet been proposed for software, including SOA, and many of the SOA issues discussed herein are direct results of this problem, as noted below.

The SOA Bandwagon

SOA has been embraced in a flood of articles, books, websites, and consultancies dedicated to promoting the approach. These all give the impression that SOA is a broadly accepted, widely implemented, and highly successful technique; however, it is very hard to find documented, published examples of organizations that are using SOA successfully. In the commercial world, SOA has certainly not yet fulfilled its grand vision of providing an open marketplace in which services that are created and published by any organization, vendor, or individual can be discovered and invoked at runtime by any other service or user on the Internet, without prior arrangement with the service's creator. If indeed SOA has been successful, it appears to have been so only within the confines of specific organizations (i.e., corporate computing environments) that utilize their own internally created services to implement their own internal business processes. This type of SOA may be built on top of an "Enterprise Service Bus" (ESB), which provides enterprise-wide messaging. However, even this kind of internal success is largely speculative, since there are few published examples of it, leading to the alternate conclusions either that SOA has not yet been successful at all, or that all of its successes have been kept secret by the organizations that have profited from them.

In summary, it appears that SOA has at best had limited success in the commercial environment, but the evidence for this is sparse: there are no widely known and publicized success stories, nor has there been much research or development of SOA protocols for micro-payment or other schemes that would be expected to be in use if a robust SOA marketplace were evolving.

Nevertheless, the idea of SOA is rapidly being accepted by many segments of the software development market, due to the combination of its universally positive publicity (hype) and its promise of providing increased interoperability at a time when virtually all segments are facing increasing challenges to consolidate, integrate, and rationalize their existing software systems.

In particular, DoD has adopted SOA as the technology underlying its NetCentric Warfare (NCW) doctrine and as the software architecture for the Global Information Grid (GIG). Furthermore, many distinct military computational domains, such as training and C4ISR, are planning to move toward SOA. As discussed below, there are continuing counter currents of movement toward the acquisition of monolithic enterprise resource planning (ERP) systems in segments such as logistics and human resources (HR), but SOA is becoming increasingly accepted across DoD, based at least in part on the—perhaps unwarranted—perception of its success in the commercial sphere.

The Use of SOA in the Military Environment

Putting aside the question of whether SOA has been successful in the commercial environment, it does seem to have considerable potential within many military computational domains. Its main attraction in these domains is its promise to increase interoperability and modularity among distinct computational services, thereby potentially reducing the effects of the proliferation of incompatible, overlapping, stovepiped systems that have plagued DoD for decades. In theory, at least, SOA can enable any service to interoperate with any other service, without the need to negotiate memoranda of agreement between the program offices that maintain those services or to fund or implement specific programming efforts to forge pairwise interfaces and connections between those services. SOA therefore has the potential to:

- Reduce the cost of connecting distinct systems to each other.
- Enable dynamic, on-the-fly linking of distinct systems.
- Enable dynamic new combinations of IT capabilities.
- Improve the modularity and consistency of IT capabilities.
- Allow more incremental evolution of Systems of Systems.
- Reduce redundancy and overlap among systems.
- Eliminate the need to maintain and run software everywhere.

Much of the impetus for moving toward SOA has come from the last two of these points, i.e., the desire to save money by consolidating existing, legacy systems, thereby eliminating redundant systems and programs and eliminating the need to maintain and run copies of software at every site that needs to use that software. Although SOA should ultimately produce these benefits, it is important to note that cost savings of this sort will be realized only after a rather substantial conversion investment has been made in creating and deploying SOA infrastructure and converting existing, legacy systems into SOA services. This latter point is frequently glossed over by assuming that existing systems can simply be "wrapped" in small pieces of code that make them behave as SOA services. In practice, however, many existing systems require more extensive revision or even redesign and reimplementation to become useful SOA services. At the very least, existing user interfaces, application programming interfaces (APIs), networking interfaces, remote procedure call interfaces, and data access mechanisms may have to be revised, redesigned, disabled, or excised from a legacy program in order to turn it into an SOA service.

More fundamentally, the capabilities provided by a given legacy system or group of systems may not be factored appropriately to work as SOA services: the functionality provided by a system or group of systems may have to be re-factored into a very different set of modular capabilities in order to produce useful SOA services. Furthermore, although it is possible to re-factor a single system in isolation this way, doing so cannot eliminate the redundancy present in a group of systems. In order to eliminate such redundancy, it may be necessary to re-factor an entire group of existing systems as a whole, which may increase the difficulty of evolving incrementally toward leaner and more effective Systems of Systems. This implies that the fifth and sixth bullets above (incremental evolution and reduction of redundancy) are at odds with one another and may to some extent be in irreducible conflict with each other.

Nevertheless, the potential advantages of SOA make it an attractive approach in many military computational domains. For example, it may be a more flexible and effective alternative to creating monolithic ERP systems for logistics, HR, and other military business functions. Moreover, it seems well suited to the creation of more modular and flexible C4ISR capabilities and a wide range of warfighter services, including and extending beyond NCW. Of particular relevance here, SOA appears to offer a number of potential advantages that might improve the scheduling, management, and oversight of training and leader development across the Army.

Of course, a computational architecture such as SOA is merely one aspect of an IT infrastructure, and even an IT infrastructure as a whole plays merely a supporting role in the design, management, and execution of training and leader development. Therefore, neither IT in general nor SOA in particular can provide a silver bullet for improving ATLD. However, moving toward SOA may help the Army consolidate and integrate its IT systems that support ATLD, thereby improving the efficiency and effectiveness of management and decision processes in this domain.

Caveats for SOA

Two major caveats for SOA are that it requires the creation of an appropriate set of services and that the semantics of these services must be understood and matched in order for an orchestrated SOA application to work correctly.

An SOA environment consists of infrastructure plus a population of services. The infrastructure provides standards and mechanisms that enable services to describe themselves, publish their descriptions, discover each other at runtime, authenticate and invoke each other, pass data back and forth between each other, and orchestrate sequences of service invocations to perform desired business processes. However, in order for this infrastructure to be useful, the SOA environment must be populated with a set of services that are appropriate for performing the range of business processes that are of interest. These services can sometimes be adapted from existing applications, but as pointed out above, this often requires re-factoring those existing applications and groups of applications into new functional pieces. Moreover, even when such re-factoring is not required, existing applications may have to be redesigned and reimplemented to turn them into useful SOA services, for example to remove interface code that enabled the original application to interact with its users, with the network, or with other programs.

The semantics issue is even more fundamental. SOA provides interoperability among services, so that they can invoke and communicate with each other. But this does not imply that what they communicate is necessarily meaningful and appropriate for the desired computation that a user has in mind. In order to ensure semantic compatibility, services must publish semantic descriptions of their interfaces, such as the names and meanings of each data element they require and produce, as well as the semantics of the computations they perform on these data. When one service discovers another that appears to offer an appropriate functional capability, it must verify that the inputs, outputs, and computations of the service to be invoked have the desired semantics. Each service's semantics must be understood by every other service that invokes it. The names and meanings of inputs and outputs of services must therefore be easy to understand and unambiguous, and their computations may have to be described in some detail. Formal semantic specification languages are provided by the SOA environment for this purpose, but using these languages to encode the semantics of a service requires considerable work, and understanding such specifications once they have been encoded requires additional work.

The principle of SOA is that services and orchestration code should be able to interpret these semantic descriptions automatically, thereby relieving the user of the burden of having to do so; but this ideal may be difficult to realize in many cases. The success of SOA therefore rests heavily on the degree to which such semantic descriptions of services can be encoded and understood by other services and ultimately by their users.

The Downside of Using SOA

In addition to the above caveats, there are some negative aspects of using SOA, including:

- Decreased autonomy, control, and (potentially) access.
- Potential loss of internal expertise.
- Potential decrease in confidence.
- Decreased usability.
- Increased network demands.

The first three of these are direct outgrowths of the advantages of SOA. Because SOA makes it possible to invoke remote services rather than running local software, it makes it unnecessary for each organization to acquire, maintain, and run its own copy of such software. This implementation of "Software as a Service" (SaaS) reduces local administrative costs and burdens, but it also decreases local control over the software and makes access to it dependent on network and remote server availability. Similarly, since less software is maintained and run locally, there may be a reduction in the internal expertise that most organizations maintain in using that software. This may lead to increased dependence on remote help desk services and administrative support. Moreover, although most organizations should reap the benefit of a reduced burden in updating software and ensuring that it complies with Information Assurance and security regulations, this comes at the cost of having to trust that the remote organizations that offer SOA services will perform these functions competently and reliably.

The last two bullets above are more unalloyed drawbacks to SOA, since they are not the reverse side of a coin of positive benefit. Due to the inherent separation of the user (or "client") layer from the remote service layer in SOA, user interaction is necessarily limited. It is difficult to make an SOA application behave as responsively or be as interactive as a standalone application that runs entirely on a user's local system. It is more difficult to provide immediate feedback using SOA, such as responding to the user's moving a mouse over an image or map or responding to an invalid keystroke as the user types a data value (before the user "submits" or "enters" the entire value field). Although such limitations can be offset to some extent by downloading and running "client-side" software on each user's machine, doing so undermines the SOA paradigm and creates maintainability and Information Assurance issues. This drawback is also present in client-server architectures, but it may be more serious in SOA, since most services are designed to be invoked by other services and cannot assume that they will be interacting directly with a client.

Finally, SOA is inherently bandwidth intensive, since it requires considerable inter-service communication and protocol overhead to execute an orchestrated business process that invokes multiple services and passes data and results among them. This increases bandwidth requirements and latency compared to standalone software approaches.

SOA Versus Traditional, Stovepiped Solutions

Most existing systems in the ATLD domain, as well as elsewhere in the Army and across DoD, were designed in an isolated, stovepiped manner. Individual organizations or subcommunities with specific needs either secured formal funding to establish a program office to develop program of record systems, or developed small-scale systems informally. These systems were often designed as "point solutions" to a specific problem and as such were often not intended to interoperate much with other systems. Isolating system design and development in stovepipes in this way has the advantage that the requirements are simpler and that design, implementation, and maintenance are not dependent on coordination with the design, implementation, and maintenance efforts of other systems. Designing a standalone system in this way is often the quickest, cheapest, and most effective way to achieve a specific IT capability, and it requires the least coordination and compromise with other systems. Within ATLD, this stovepiped approach has produced a number of relatively isolated systems, each of which deals with an isolated aspect of training and leader development, such as programming, funding, developing, managing, scheduling, and executing courses, as well as allocating, managing, and scheduling the instructors, facilities, equipment, and IT capabilities needed to deliver training.

As such point solutions have proliferated, it has become apparent that many decisions are suboptimized by focusing narrowly on one aspect of the overall ATLD enterprise, making it difficult to ensure that the results are optimum for the ATLD enterprise and, moreover, for the Army enterprise as a whole. Isolation of IT systems from each other is merely one aspect of this problem, since the stovepiping often extends as well to the organizations that use these systems. However, the difficulty of integrating and interoperating among ATLD systems makes it harder to generate and view the enterprise perspective that is now widely sought.

To address this lack of interoperability, individual IT systems are often connected to each other on a "pairwise" basis, by forging a specific interface between them for a specific purpose. However, such pairwise efforts often require formal agreement between program offices, allocation of funding and manpower resources, and focused coding effort to create a new interface. Even then, the resulting interoperability is often limited to a specific need, so that expanding that interoperation in the future requires additional work or even a new pairwise interface.

Although the point solution approach to system design and the associated pairwise interface approach to interoperability work up to a point, they do not scale well. Each new pairwise interface that a system adds complicates that system with code that is specific to some other system and creates new dependencies between those systems. These dependencies add to the maintenance burden for the system and make it harder to redesign or improve the system without risking the disruption of its interfaces to other systems. Furthermore, any change to one of these other systems may break its pairwise interfaces, requiring all systems to which it is interfaced to repair or redesign their interfaces with it.

As recognition of the importance of an enterprise view of training and leader development within the broader Army enterprise grows, it has become apparent that a more scalable approach to interoperability would improve the reliability, maintainability, extensibility, and effectiveness of the connections between ATLD systems, resulting in better integration of training-related information that should improve decision making. Two of the most popular current approaches to achieving such increased integration of information are enterprise resource planning and service-oriented architecture, which are discussed in the following sections.

SOA Versus ERP

Most large-scale enterprises have evolved large numbers of distinct IT systems and data enclaves corresponding to their organizational, functional, or geographic units, offices, and departments. These systems have often been developed and evolved separately to serve the specific needs of the units or groups that have commissioned them and so tend to be stovepiped. Even when IT departments develop systems for their enterprises, the resulting systems tend to have limited scope, corresponding to the specific functions of the groups for which they are developed. In addition, many systems are developed "beneath the radar" by local units or groups for their own purposes, without regard to the overall needs of the enterprise. These systems are often built using simple tools, such as spreadsheet macro languages, resulting in solutions that do not scale well. Furthermore, these local systems are often lashed together in an ad hoc fashion, sometimes long after their initial development, without regard to interoperability, scalability, information assurance concerns, or recognized IT system design principles. This produces systems that do not communicate with each other or that maintain data in different formats or with different semantics, making it difficult to integrate their functions across the enterprise.

ERP is an attempt to integrate all of the data and IT systems of an enterprise, in order to improve efficiency, effectiveness, management, and decision making processes. A typical ERP solution tends to be a single, monolithic system that attempts to do everything an enterprise needs, maintaining a single, integrated database containing all of the enterprise's data. Because such monolithic systems are very large and therefore difficult to customize or modify, software vendors have attempted to develop generic, one-size-fits-all ERP products, which have proved to be a poor fit for many enterprises. An organization that acquires or develops such an ERP typically faces an all-or-nothing choice: either use the ERP for everything or abandon it. If an ERP's data structures or business processes fail to fit those of a particular organization, the organization's choices are:

1. Change its processes to fit the ERP.
2. Write custom code to modify the ERP.
3. Pay the ERP vendor to write such custom code.
4. Give up on the ERP and return to using legacy systems.

Yet changing an enterprise's business processes may undermine its organizational culture or even its core capabilities, whereas writing custom code is expensive and problematic, since whenever a new version of the ERP is released, such custom code is likely to break and require revision. Furthermore, even custom code may have a limited ability to adapt a given ERP to an enterprise's business model.

In contrast, as pointed out above, SOA is a computational architecture. As such, it is an environment, not a system like an ERP, nor is it an approach to performing a specific set of enterprise business processes, such as payroll, account management, customer relations, or manufacturing. SOA provides an alternative to ERP for performing enterprise functions. Whereas ERP adopts a monolithic, single-system approach, SOA encourages a more modular, flexible approach. Services can be implemented incrementally and simply plugged together as needed to perform desired business processes. If a new business process requires some functional capability that is missing, a new service can be created to provide this capability and can be made available in the SOA environment. Old services can be replaced with new, improved ones, with little or no impact on other services or on the orchestration programs that define particular business processes. Customization can be performed by modifying orchestration programs or by modifying or replacing specific services, without (in principle) having to change any other component of the environment.

Experiences with ERP to date have been quite problematic. Many military and commercial organizations have spent huge amounts of time, effort, and money on ERPs only to find that these systems were ultimately too inflexible and difficult to use and maintain or that using them required unacceptable adaptations by the organization.

However, the goal of integrating enterprise data and IT systems to improve enterprise performance remains a highly worthwhile one, to which SOA may offer an alternative that avoids many of ERP's problems. SOA itself is still quite immature and largely unproven, yet it is so far the most promising approach available to improve interoperability and flexibility across large sets of IT components. It therefore seems likely that SOA could be used effectively to incrementally implement a "modular ERP-like capability" for ATLD that would offer the potential benefits of ERP without its major limitations of inflexibility and monolithic, all-or-nothing acquisition and implementation.

In summary, the difficulty of integrating and interoperating among ATLD systems makes it harder to generate and view the enterprise perspective that is increasingly sought within the Army. As recognition of the importance of an enterprise view of training and leader development has grown, it has become apparent that a more scalable approach to interoperability would improve the reliability, maintainability, extensibility, and effectiveness of the connections between ATLD systems, resulting in better integration of training-related information that should improve decision making.

SOA is the most promising current approach to addressing this lack of interoperability and is, moreover, being aggressively pursued throughout the net-centric operational environment, as well as within ATLD's ATIA and ATIS efforts. SOA offers a more scalable approach than traditional pairwise agreements and interface development involving specific IT systems. At the same time, it avoids the all-or-nothing risk and operational cost of an ERP approach, which would attempt to fully integrate ATLD capabilities by replacing all existing ATLD systems with a single new system. Furthermore, SOA is particularly attractive as a way of sharing training-related information among systems, since this could be done by having each relevant system develop the relatively small new SOA services needed to exchange such information without having to convert the bulk of each system to SOA. This approach—if coupled with suitable new instrumentation efforts as described in the report—offers a relatively low-cost means of sharing and analyzing training information across ATLD.

Bibliography

Ackerman, Philip L., "Individual Differences in Skill Learning: An Integration of Psychometric and Information Processing Perspectives," *Psychological Bulletin*, Vol. 102, 1987, pp. 3–27.

Agarwal, Ritu, and Jayesh Prasad, "The Antecedents and Consequents of User Perceptions in Information Technology Adoption," *Decision Support Systems*, Vol. 22, No. 1, 1998, pp. 15–29.

Alliger, George, Scott I. Tannenbaum, Winston Bennett, Holly Traver, and Allison Shotland, "A Meta-Analysis of Relations Among Training Criteria," *Personnel Psychology*, Vol. 50, 1997, pp. 341–358.

Alvarez, Kaye, Eduardo Salas, and Christina M. Garofano, "An Integrated Model of Training Evaluation and Effectiveness," *Human Resource Development Review*, Vol. 3, No. 4, 2004, pp. 385–416.

Bernard, Robert M., Philip C. Abrami, Yiping Lou, Evgueni Borokhovski, Anne Wade, Lori Wozney, Peter A. Wallet, Manon Fiset, and Binru Huang, "How Does Distance Education Compare with Classroom Instruction? A Meta-analysis of the Empirical Literature," *Review of Educational Research*, Vol. 74, No. 3, 2004, pp. 379–439.

Bloom, Benjamin S., Max D. Englehart, Edward J. Furst, Walker H. Hill, and David R. Krathwohl, *Taxonomy of Educational Objectives: The Classification of Educational Goals. Handbook I: Cognitive Domain*. New York, Toronto: Longman, 1956.

Brown, Kenneth G., "Examining the Structure and Nomological Network of Training Reactions: A Closer Look at 'Smile Sheets,'" *Journal of Applied Psychology*, Vol. 90, 2005, pp. 991–1001.

Carey, N., D. L. Reese, D. F. Lopez, R. W. Shuford, and J. K. Wills, *Time to Train in Self-Paced Courses and the Return on Investment from Course Conversion*, Alexandria, Va.: The CNA Corporation, CRM D0015039.A1/SR1, 2007.

Dweck, C. S., "Motivational Processes Affecting Learning," *American Psychologist*, Vol. 41, 1986, pp. 1040–1048.

Ely, Katherine, Traci Sitzmann, G. A. Redding, and Cari Falkiewicz, *Does Additional Time in Online Training Result in Higher Learning Outcomes for Electronics Technicians?* Washington, D.C.: Advanced Distributed Learning Initiative, Technical Report 2008-1, 2008.

Fisher, Sandra L., Michael E. Wasserman, and Karin A. Orvis, "Enhancing E-learning Effectiveness Through Learner Engagement," paper presented at the 19th Annual Conference of the Society for Industrial and Organizational Psychology, Chicago, Ill., April 2004.

Gagne, R. M., "Learning Outcomes and Their Effects: Useful Categories of Human Performance," *American Psychologist*, Vol. 39, 1984, pp. 377–385.

Goldstein, Irwin L., and J. Kevin Ford, *Training in Organizations,* Belmont, Calif.: Wadsworth, 2002.

Hamilton, Laura S., Stephen P. Klein, and William Lorie, *Using Web Based Testing for Large-Scale Assessment*, Santa Monica, Calif.: RAND Corporation, IP-196, 2000.
http://www.rand.org/pubs/issue_papers/IP196.html

Holton, E. F., III, "The Flawed Four-Level Evaluation Model," *Human Resource Development Quarterly*, Vol. 7, 1996, pp. 5–21.

Kirkpatrick, Donald L., *Evaluating Training Programs: The Four Levels,* San Francisco: Berrett-Koehler, 1994.

Klein, Howard J., Raymond A. Noe, and Chongwei Wang, "Motivation to Learn and Course Outcomes: The Impact of Delivery Mode, Learning Goal Orientation, and Perceived Barriers and Enablers," *Personnel Psychology*, Vol. 59, 2006, pp. 665–702.

Kraiger, Kurt, "Decision-Based Evaluation," in K. Kraiger (ed.), *Creating, Implementing, and Managing Effective Training and Development*, San Francisco, Calif.: Jossey-Bass, 2002, pp. 331–375.

Kraiger, Kurt, J. Kevin Ford, and Eduardo Salas, "Application of Cognitive, Skill-Based, and Affective Theories of Learning Outcomes to New Methods of Training Evaluation," *Journal of Applied Psychology*, Vol. 78, 1993, pp. 311–328.

Lawther, Peter M., and Derek H. T. Walker, "An Evaluation of a Distributed Learning System," *Education & Training*, Vol. 43, No. 2/3, 2001, pp. 105–116.

Lewis, Eileen L., and Elaine Seymour, "Attitude Surveys," Field-Tested Learning Assessment, web page, undated. As of April 13, 2010:
http://www.flaguide.org/cat/attitude/attitude7.php
PDF version is available. As of February 2011:
http://www.flaguide.org/extra/download/cat/attitude/attitude.pdf

Lord, Frederic M., *Applications of Item Response Theory to Practical Testing Problems,* Mahwah, N.J.: Lawrence Erlbaum Associates, Inc., 1980.

Mesmer-Magnus, Jessica, and Chockalingam Viswesvaran, "Inducing Maximal Versus Typical Learning Through the Provision of a Pretraining Goal Orientation," *Human Performance*, Vol. 30, No. 3, 2007, pp. 205–222.

Morey, John C., Michael D. Bush, Robert Beebe, Scott McPhail, and William R. Bickley, *Best Practices for Using Mobile Training Teams to Deliver Noncommissioned Officer Education Courses*, Andover, Mass.: Dynamics Research Corporation, 2009.

Mumford, Michael D., David P. Costanza, Wayne A. Baughman, Victoria K. Threlfall, and Edwin A. Fleishman, "Influence of Abilities on Performance During Practice: Effects of Massed and Distributed Practice," *Journal of Educational Psychology*, Vol. 86, No. 1, 1994, pp. 134–144.

Paradise, Andrew, and Lahel Patel, *State of the Industry Report*, Alexandria, Va: American Society for Training & Development, 2009.

Peterson's, a Nelnet Company, "What Kind of Learning Environment Would Be Best for You: A Traditional Setting or a Distance Learning Program?" Web page, undated. As of April 14, 2009:
http://www.petersons.com/dlwizard/code/default.asp

Salas, Edward, and Janice A. Cannon-Bowers, "The Science of Training: A Decade of Progress," *Annual Review of Psychology*, Vol. 52, 2001, pp. 471–499.

Schonlau, Matthias, Ronald D. Fricker, and Marc N. Elliott, *Conducting Research Surveys Via E-Mail and the Web*, Santa Monica, Calif.: RAND Corporation, MR-1480-RC, 2002.
http://www.rand.org/pubs/monograph_reports/MR1480.html

Shanley, Michael G., James C. Crowley, Matthew W. Lewis, Susan Straus, Kristin Leuschner, and John Coombs, *Making Improvements to the Army Distributed Learning Program*, Santa Monica, Calif.: RAND Corporation, MG-1016-A, forthcoming.

Sitzmann, Traci, Kenneth G. Brown, Wendy J. Casper, Katherine Ely, and Ryan D. Zimmerman, "A Review and Meta-Analysis of the Nomological Network of Trainee Reactions," *Journal of Applied Psychology*, Vol. 93, 2008, pp. 280–295.

Sitzmann, Traci, Kurt Kraiger, Dennis W. Stewart, and Robert A. Wisher, "The Comparative Effectiveness of Web-Based and Classroom Instruction: A Meta-Analysis," *Personnel Psychology*, Vol. 59, 2006, pp. 623–664.

Straus, Susan G., Michael G. Shanley, Rachel M. Burns, Anisah Waite, and James C. Crowley, *Improving the Army's Assessment of Interactive Multimedia Instruction Courseware*, Santa Monica, Calif.: RAND Corporation, MG-865-A, 2009.
http://www.rand.org/pubs/monographs/MG865.html

Tannenbaum, Scott I., Janis A. Cannon-Bowers, Eduardo Salas, and John E. Mathieu, *Factors That Influence Training Effectiveness: A Conceptual Model and Longitudinal Analysis*, Orlando, Fla.: Naval Training Systems Center, Technical Rep. No. 93-011, 1993.

Training and Doctrine Command (TRADOC), *Systems Approach to Training: Course and Courseware Evaluation*. Fort Monroe, Va.: Headquarters, Department of the Army, United States Army Training and Doctrine Command, 350-70-10, 2004.

Training and Doctrine Command (TRADOC), *The United States Army Learning Concept for 2015*. Fort Monroe, Va.: Headquarters, Department of the Army, United States Army Training and Doctrine Command, 525-8-2, 2011.

University of Wisconsin-Whitewater, "Online Course Readiness Assessment," website, 2006. As of January 2011:
http://www.uww.edu/icit/olr/stu/gettingstarted/manual.html.

Wang, Morgan C., Charles D. Dziuban, and Patsy D. Moskal, "A Web-Based Survey System for Distributed Learning Impact Evaluation," *The Internet and Higher Education*, Vol. 2, No. 4, 2000, pp. 211–220.